Computational Intelligence Application in Electrical Engineering

Computational Intelligence Application in Electrical Engineering

Editors

Marinko Barukčić
Nebojša Raičević
Vasilija Šarac

MDPI • Basel • Beijing • Wuhan • Barcelona • Belgrade • Manchester • Tokyo • Cluj • Tianjin

Editors
Marinko Barukčić
Faculty of Electrical Engineering, Computer Science and Information Technology Osijek, J.J.Strossmayer University Osijek, Trpimirova 2B, 31 000 Osijek, Croatia

Nebojša Raičević
Faculty of Electronic Engineering, University of Niš, Aleksandra Medvedeva 14, 18000 Niš, Serbia

Vasilija Šarac
Faculty of Electrical Engineering, University Goce Delchev, 2000 Štip, North Macedonia

Editorial Office
MDPI
St. Alban-Anlage 66
4052 Basel, Switzerland

This is a reprint of articles from the Special Issue published online in the open access journal *Electronics* (ISSN 2079-9292) (available at: https://www.mdpi.com/journal/electronics/special_issues/CIA_electronics).

For citation purposes, cite each article independently as indicated on the article page online and as indicated below:

LastName, A.A.; LastName, B.B.; LastName, C.C. Article Title. *Journal Name* **Year**, *Volume Number*, Page Range.

ISBN 978-3-0365-4695-7 (Hbk)
ISBN 978-3-0365-4696-4 (PDF)

Contents

About the Editors

Marinko Barukčić

Marinko Barukčić (Professor) earned his BsC, MsC, and PhD degrees in 1998, 2008, and 2012, respectively, from the Faculty of Electrical Engineering, Computer Science, and Information Technology in Osijek, Croatia. From 1998 to 2007, he worked at the Croatian Electricity Distribution System Operator. From 2007 to date, he has been working at the Faculty of Electrical Engineering, Computer Science and Information Technology in Osijek. Marinko is involved in several scientific projects as a project leader and member of a project team. At the faculty, he lectures in various courses for Bachelor, Master and PhD degree programs. He is the author of more than 80 scientific articles in scientific journals and conferences. His scientific interests include modeling, analysis and the simulation of power systems, electrical machines and computation of electromagnetic fields. He is particularly interested in the application of computational intelligence techniques in electrical engineering.

Nebojša Raičević

Nebojša Raicević (Professor) was born in Nis, Serbia, 1965. He received BEng, MSc and magistrand degrees from University of Nis, in Faculty of Electronic Engineering in Nis. He defended his PhD thesis, entitled "Calculation of electric field in cable accessories", at the Department of Theoretical Electrical Engineering, University of Nis. He is a Full Professor in the Faculty of Electronic Engineering, author of new numerical method (Hybrid Boundary Element Method (HBEM)) and he has published more than 200 papers in international journals or conference proceedings and served as a project manager in research in many national and international projects. He was a mentor of several doctoral dissertations and the President of the IEEE EMC Section for Serbia and Montenegro. His research interests include numerical methods for electromagnetic problem-solving, power cable terminations and joints, partial discharges, grounding electrodes systems, non-linear electrostatic problems, anisotropic materials and metamaterials application determination, magnetic field calculation of coils and permanent magnets.

Vasilija Šarac

Vasilija Šarac (Professor) received BsC, MsC, and PhD degrees in 1995, 1999, and 2005, respectively, in the Faculty of Electrical Engineering, University "Ss. Cyril and Methodius" Skopje. She is a Full Professor in the University "Goce Delcev" Shtip, Faculty of Electrical Engineering. From 1995 to 2010, she worked with Siemens A.E. Representative office, Skopje. Vasilija collaborated scientifically and professionally with some international institutions and research institutes and participated in training in international research institutions and companies. She has been involved in several projects and held the position of vice-dean in her faculty. She is a member of professional organizations and was a member of a number of scientific committees of conferences. Vasilija has a scientific interest in electrical machines, power converters, circuit analysis and electromagnetic compatibility, especially in optimizing the design and control of electrical machines. She published about 150 scientific papers in scientific journals and conferences.

 MDPI

Editorial

Computational Intelligence Application in Electrical Engineering

Marinko Barukčić [1,*], Vasilija Šarac [2] and Nebojša Raičević [3]

1 Faculty of Electrical Engineering, Computer Science and Information Technology Osijek, Josip Juraj Strossmayer University of Osijek, 31000 Osijek, Croatia
2 Faculty of Electrical Engineering, University Goce Delchev, 2000 Stip, North Macedonia; vasilija.sarac@ugd.edu.mk
3 Faculty of Electronic Engineering, University of Nis, Aleksandra Medvedeva 14, 18000 Nis, Serbia; Nebojsa.Raicevic@elfak.ni.ac.rs
* Correspondence: marinko.barukcic@ferit.hr; Tel.: +385-31-224-685

Citation: Barukčić, M.; Šarac, V.; Raičević, N. Computational Intelligence Application in Electrical Engineering. *Electronics* **2022**, *11*, 1883. https://doi.org/10.3390/electronics11121883

Received: 10 June 2022
Accepted: 14 June 2022
Published: 15 June 2022

Nowadays, scientists and practitioners in the field of electrical engineering observe the increasing application of information technology, computers, and computing techniques. Modern concepts such as smart power grids and smart industries require a multidisciplinary approach and a close connection and synergistic application of IT and computer hardware and software in all areas of electrical engineering. In addition, the application of advanced computational tools is essential for the simulation and modeling of complex electrical systems and devices. The application of computational tools based on numerical mathematical methods has enabled practical calculations in the field of electromagnetic field theory with more realistic models of the devices.

The Special Issue "Computational Intelligence Application in Electrical Engineering" aims to promote the techniques and procedures of computational intelligence for modeling, optimization, simulation, and computation in various fields of electrical engineering. Thanks to the authors' interest in this Special Issue, seven research and review articles were published out of the ten submitted papers.

In the review article [1], the authors provide an overview of the application of computational intelligence methods in power engineering, in particular, the application of computational intelligence in the field of power grids. The article addresses various goals of applying computational intelligence in the area of smart power grids, such as optimal scheduling of distributed generation and optimization of smart power grid management. The remaining published articles are original research papers.

The authors of [2] have proposed a co-simulation approach to solve the very complex optimization problem of optimal allocation of distributed generation assets and power control of controllable distributed generation assets. The optimization problem is of black-box type, and an artificial neural network is proposed for the distribution of the output power of the distributed generation units.

Paper [3] presents the application of a metaheuristic optimization method for optimal coordination of directional overcurrent relays and distance relays in the second zone. The authors developed a modified school-based optimization method as an improvement to the basic version of this optimizer.

In [4], the uncertainty in the scheduling of electricity distribution generation is presented considering the electricity market. The modeling and impact of different uncertainties (in the intensity of primary energy sources as well as in the energy price) at the intraday market level was developed and proposed here.

The improved method for power flow calculations in power systems was developed in [5]. The proposed method uses the Newton- S-Iteration Process and shows advantages over classical power flow methods especially for ill-conditioned systems.

The authors of [6] have developed a procedure for the analysis and optimization of a synchronous motor with line start and asymmetric permanent magnet arrangement in the rotor. The method includes a finite element analysis of the motor using a simulation program. The optimization method is based on a parametric analysis of the steady state and transients of the motor.

The last published article (in chronological order) [7] in the Special Issue deals with the optimal control of induction motors. In the article, the application of a fuzzy controller for the predictive current control of an induction motor was developed. It also presents the optimization of the parameters of the fuzzy controller using a co-simulation approach and a metaheuristic optimization method.

We would like to thank all the authors for their interest and contributions to this Special Issue. We thank the reviewers who contributed to the quality of the presentation of the articles with their constructive comments and suggestions. We thank the editorial board of the journal *Electronics* for the invitation and the opportunity to edit this special issue. A big thank to the editorial board for taking care of the whole process and making sure that everything was ready on time. Our special thanks to Ms. Hebbe Tian, the assistant editor of the Special Issue, for her kindness and timely completion of all the steps in this Special Issue.

Funding: This research received no external funding

Conflicts of Interest: The authors declare no conflict of interest.

References

1. Vukobratović, M.; Marić, P.; Horvat, G.; Balkić, Z.; Sučić, S. A Survey on Computational Intelligence Applications in Distribution Network Optimization. *Electronics* **2021**, *10*, 1247. [CrossRef]
2. Barukčić, M.; Varga, T.; Štil, V.J.; Benšić, T. Co-Simulation Framework for Optimal Allocation and Power Management of DGs in Power Distribution Networks Based on Computational Intelligence Techniques. *Electronics* **2021**, *10*, 1648. [CrossRef]
3. Abdelhamid, M.; Kamel, S.; Korashy, A.; Tostado-Véliz, M.; Banakhr, F.A.; Mosaad, M.I. An Adaptive Protection Scheme for Coordination of Distance and Directional Overcurrent Relays in Distribution Systems Based on a Modified School-Based Optimizer. *Electronics* **2021**, *10*, 2628. [CrossRef]
4. Garcia-Guarin, J.; Alvarez, D.; Rivera, S. Uncertainty Costs Optimization of Residential Solar Generators Considering Intraday Markets. *Electronics* **2021**, *10*, 2826. [CrossRef]
5. Tostado-Véliz, M.; Kamel, S.; Taha, I.B.M.; Jurado, F. Exploiting the S-Iteration Process for Solving Power Flow Problems: Novel Algorithms and Comprehensive Analysis. *Electronics* **2021**, *10*, 3011. [CrossRef]
6. Sarac, V.; Minovski, D.; Janiga, P. Parametric Analysis for Performance Optimization of Line-Start Synchronous Motor with Interior Asymmetric Permanent Magnet Array Rotor Topology. *Electronics* **2022**, *11*, 531. [CrossRef]
7. Varga, T.; Benšić, T.; Barukčić, M.; Štil, V.J. Optimization of Fuzzy Controller for Predictive Current Control of Induction Machine. *Electronics* **2022**, *11*, 1553. [CrossRef]

Review

A Survey on Computational Intelligence Applications in Distribution Network Optimization

Marko Vukobratović [1,*], Predrag Marić [2], Goran Horvat [1], Zoran Balkić [1] and Stjepan Sučić [3]

1 Base58 Ltd., 31000 Osijek, Croatia; goran.horvat@base58.hr (G.H.); zoran.balkic@base58.hr (Z.B.)
2 Faculty of Electrical Engineering, Computer Science and Information Technology, 31000 Osijek, Croatia; predrag.maric@ferit.hr
3 Končar—Power Plant and Electric Traction Engineering Inc., 10000 Zagreb, Croatia; stjepan.sucic@koncar-ket.hr
* Correspondence: marko.vukobratovic@base58.hr

Abstract: This paper aims to present carefully selected scientific papers that have pushed the boundaries in the application of advanced computational intelligence–based methods in power engineering, mainly in optimal power system management. Contemporary development of the Smart Grid and detailed framework for power grid digitalization enabled the real and efficient application of advanced optimization algorithms presented in this paper. Papers that are not directly related to Smart Grid management are also considered, since they solve the partial challenges of planning and development with metaheuristic procedures, and according to the authors, they are highly applicable and represent a fundamental starting point for wider application. This paper covers papers and research whose results are reproducible and can be realized in production-grade software. The emphasis of the paper is on the considerate and impartial way of providing a concise overview of the methods for solving technical challenges within the accepted Smart Grid architecture. The paper is the result of many years of research and commitment to this field and represents the foundation for present research and development.

Keywords: active distribution network; computational intelligence; optimization algorithms; optimal distribution system management; optimal Smart Grid management; advanced distribution system optimization; renewable distributed generation; Smart Grid optimization

Citation: Vukobratović, M.; Marić, P.; Horvat, G.; Balkić, Z.; Sučić, S. A Survey on Computational Intelligence Applications in Distribution Network Optimization. *Electronics* **2021**, *10*, 1247. https://doi.org/10.3390/electronics10111247

Academic Editor: Osvaldo Gervasi

Received: 1 May 2021
Accepted: 20 May 2021
Published: 24 May 2021

1. Introduction

Smart Grid research is the new, challenging area with great interest of research teams in the EU space [1,2]. Future advanced and smarter power network relies heavily on the possibility of independent permanence and self-sufficiency based on data and informed management [3,4]. The development of the power network at the distribution level and the increasing integration of renewable energy sources (RES) makes the network heterogeneous and more diverse. The laws of network management that were once applicable to the whole system become less valid in the era of changing power system and novel paradigms. Such system requires advanced methods of rapid analysis of a multitude of possible scenarios to achieve optimal power system management. This paper examines the scientific and engineering foundations of using computational intelligence (CI) methods for achieving optimal real-time or quasi-real time management of Active Distribution Network (ADN). ADN is the first stage in the formation of more advanced Smart Grid which involves usual power distribution equipment along with information and communication technologies (ICT) and monitoring systems. The goal of Smart Grid is more reliable power supply and adequate power quality (PQ) in the ever-changing environment. The presented methods are currently oblivious to information technology and are focusing on electrical engineering principles since those are the foundations for every other application. Objectives observed by this paper are energy balancing, losses reduction, reliability increase and policy definitions described in latest scientific achievements. Energy balancing is the most important part of Smart Grid and it can be a topic for itself, especially when market considerations become equally important to technical ones.

Although most scientific papers presented here focused on optimal Distributed Generation (DG) allocation and sizing in distribution network, objectives achieved by papers presented in this work can be used as an engineering foundation for ADN operational management based on advanced CI optimization algorithms. The scope covered by Distribution Network Optimization is much larger than just the allocation of DG units and includes optimization of the topology and timetable planning of existing power plants to achieve optimal power flows [5,6]. Accordingly, this paper includes the papers that in a concise and precise way define the application of CI methods in solving modern challenges in the power system, especially the problem of utilization of a larger number of DG units. Real-time power flow optimization, as described Reference [5], comes only after a real-world software solution is set up at the planning level. This paper defines the principles that such a software solution should meet and describes the leading CI algorithms that can result in such a software solution. Subsequently, a new part of such a solution may be real-time optimization module for optimization of power flows, voltage profile and oscillation damping control, as given in Reference [5]. For such real-time power flow optimization with limited infrastructure for data acquisition in the distribution grid, a robust state estimation engine is crucial [6] to address uncertainties in the distribution network.

Scientific works observed by this paper can lead to the development of ADN management solutions and the current activities of the authors confirm that hypothesis. Assumptions of ADN operational management by CI is unequivocally and unanimously highlighted and validated by papers [7–9]. The challenges ahead in the field of Big Data collection and consumption monitoring using advanced metering infrastructure are certainly part of the development of such ADN management solutions [6].

The rest of the paper is structured as follows: Section 2 provides the overview of the used methodology, sources of information, knowledge collection, synthesis and solution assessment criteria. Section 3 analyzes the scope of DG impact in the distribution network and what are physical constraints that need to be respected when evaluating applicability of any CI method. Section 3 presents most important papers that use CI for optimal allocation and optimal scheduling of DG units, while considering the physical and technical limitations of the considered ADN system. Selected papers are the result of described methodology and selection criteria. Moreover, by the opinion of the authors of this paper, selected papers can be utilized in market-oriented planning and scheduling scenarios in real-world systems nowadays. Section 4 describes papers that bring significant breakthroughs in the operational management of the ADN with multiple DG units. Finally, the conclusion outlines observations and knowledge gathered by this paper.

2. Methodology

The papers covered in this paper are partly part of the research during the preparation of the doctoral dissertation, and partly during the research phase of the project mentioned in the acknowledgment. Given the constant changes in technology, the criteria for including and excluding processed papers required fine-tuning throughout the research process. What is common to all the presented works is that the team of development experts, to a greater or lesser extent, managed to replicate the outcomes at the level of the model and the prototype solution. This was also a key criterion for the processing of individual work—high reliability that the proposed solution can be refined and developed for the needs of the actual system.

For the purpose of collecting valid information, journals indexed in the leading databases Web of Science, Scopus and Inspec were consulted. The advanced search considered more than 7000 papers. This huge amount of information has been reduced to a little over 1000 basic papers by thorough processing according to the thematic criteria and observed challenges. Additionally, by the criteria of the number of mentions and the context in which they were mentioned, and a preliminary reading of a team of researchers, the selection was reduced to around 250 relevant papers. The additional thematic division and organization of a wider team of researchers quickly prototyped the relevant group and the choice remained on those papers mentioned in this paper. If the reader considers that some paper has been unfairly neglected, the authors are open to contact and correction in future papers.

Aware of the fact that the impossibility of repeating the outcome of a single paper does not immediately imply the unacceptability of the observed solution for processing in this paper, we consulted external domain experts. The covered domains are the field of artificial and computational intelligence, distribution network optimization techniques and development of monitoring and control solutions for distribution power networks. Such an interdisciplinary team provided a multi-perspective review of the observed solutions that was in some cases crucial to understanding what was presented.

Although researchers were given the opportunity to independently assess and organize individual development teams, it turned out that regular synchronization meetings could achieve more results in less time. By dynamically adapting the research team according to Agile principles [10,11], a larger amount of methods than originally conceived was processed.

The risk of bias and false applicability estimation was solved by following the next procedure:

1. Extraction of the presented mathematical model—In this step, a multi-member team took responsibility for extracting key information from each paper and, if necessary, filling in the gaps in the mathematical model.
2. Synthesis of a defined mathematical model—If key parameters are missing in the first step, the paper is marked as incomplete and is temporarily not considered. However, it happened that some other work was based on the same laws as the one that was neglected, so one complements the other. In this step it is crucial to obtain the same or approximately the same starting point as the original authors.
3. Development of an identical or similar solution—In this step, it has proved crucial to have a team of specialized experts and profiled researchers who can understand the principles to be set out in development in order to achieve greatest effect. Usually, rapid prototyping does not require more than a few days of work for a well-organized team, and as these are very similar methods based on common principles the testing time decreased exponentially throughout the research phase.
4. Analysis of the obtained results—In this phase, the involvement of external domain experts proved to be crucial, who could intuitively, based on many years of experience, suggest measures for improvement if the results did not correspond to expectations. Moreover, they were able to explain some unexpected results and thus immediately conclude where the shortcomings of the basic observed model of some papers are. Analysis was conducted ad-hoc by the whole research team and most of the time heterogeneity of results was discussed.
5. Brainstorming session—This is a key part of the whole process; in this step, there was an exchange of ideas, knowledge and solutions, and often with loud commenting a certain hypothesis or assumption was fiercely defended. After this session, the certainty of validity was ensured and the whole process was repeated as many times as necessary to process all the papers selected for processing.

Conclusions after the comprehensive review of papers are that there are scientific databases in which papers are generally published with more detail with traceable results, while at the same time there are scientific databases that nurture papers in a shorter reporting form which makes repeating the outcomes more challenging. For some papers, mathematical experts with deep knowledge of statistics had to be consulted, and for the purposes of mathematical synthesis, several mathematical experts were engaged.

The limitations of this methodology are certainly relying on a group of researchers without using any of the artificial intelligence (AI) methods. However, the authors are of the opinion that, given the specificity of the observed challenge and previous experiences, there is greater reliability if we are guided by the opinions of domain experts and long-term scientists, than by the algorithm. In the case of considering many scientific papers for development purposes, it is more reasonable to rely on natural intelligence.

Of course, it may seem that this is completely contradictory to the topic of this paper, but it should be considered that this paper observes applications of CI, a subset of AI that always relies on exact mathematical calculations and it is in fact specifically modeled by natural intelligence.

Within the space–time possibilities and with the available literature, this paper aims to shorten the cognitive process and provide an overview in one place for all those who are just starting or are currently dealing with the field of distribution network optimization using CI methods. For better understanding, the abbreviation list is provided in the Appendix A at the end of this paper. Of course, given the context in which this research is conducted, all papers are viewed from the point of view that they must be feasible as actual optimization systems or at least as a basis for the development of future management procedures. With this work, science is viewed as a platform for launching innovative engineering solutions that should serve all of humanity.

3. DG and the Distribution Network

DG impact on the power grid is a complex mathematical problem, changing over time and depending on the parameters of each grid. Precisely because of the diversity of the system in which DG is integrated, there is no universal solution that can be magically applied to every case while respecting the laws of calculation such as real power balance equations, Jacobian Matrices and Newton—Raphson Power flow.

Current growing demand for energy in the world accompanied by an increase in energy prices, accelerated the technology development for RES utilization, which introduce additional variability. According to installed capacity, Viral et al. [12] classify RES and non-RES DG, such as micro DG with installed capacity up to 5 kW, small DG with installed capacity up to 5 MW, medium DG with installed capacity up to 50 MW and large DG with installed capacity from 50 to 300 MW. The same authors refer to the utilization of DG for the base and the peak load demand and capability to provide ancillary services to the system, while considering a power plant a DG only when it is connected to the distribution network. Similar findings on different types of DG technology and power can be found in Reference [13] in which the authors also refer to Viral et al. [12]. Sambaiah's [13] considerations of the method of optimal allocation are given in later sections. Although Reference [13] can be considered similar in its content, this paper is about reconstructed models and tested systems and is not a mere enumeration of what was read, but about presented scientific papers that the development team successfully reconstructed according to the guidelines for software development.

Considering integration location, DG is integrated in two ways: local level and end-point level [12]. Local DG means independent production units in the distribution network, and the end-point DG is a production unit integrated with the consumer. According to power system management, DG based on technologies independent of the energy resource incidence are fundamental generation while solar and wind power plants with volatile production cannot be considered as fundamental generation in power system [14,15]. However, such systems can be enriched with an energy storage and in that case increase the efficiency of the distribution network and enable the balancing of part or all of the system [16].

When modeling and developing a specific CI method for ADN management, there is a big difference depending on which type of DG integration CI needs to be observed. Aggregating end-point production at the local level has proven to be a good entry point for a challenge formulation that CI can solve more successfully.

Technical advantages of DG can relate to a wide range of influences such as the power supply of the system peak load, voltage profile improvement, system losses reduction, power supply continuity, system reliability improvement and elimination of power supply quality issues [17–19].

Voltage dips are still a challenge in many distribution grids so Ipinnimo et al. [20] point on mutual coordination of a large number of DG units in order to reduce the voltage dips. Location and power of DG are two key factors in reducing system losses [12]. Voltage dips reduction and transient voltage reductions between 10% and 90% of the nominal effective root-mean square value lasting between half cycle and one minute [9] are recognized as a key issue in distribution network with high DG penetration level, as can be seen in References [20,21]. However, there are specific circumstances in which an optimization algorithm needs to be adapted, such as a Long Distribution environment that includes manufacturing, distribution centers, terminals, geographic units, suppliers and end users [22]. Review paper of Djafar et al. hits a very specific niche and it is important to emphasize it because it can be useful to those who are in a similar environment of complex power distribution systems, without being discouraged that perhaps such a case has not been processed in the scientific literature.

Common challenges of DG integration to distribution grid mentioned by the authors of References [21,23–27] are impossibility of reactive power production of some DG technologies, necessity of power system protection settings change, possible occurrence of DG island operation, high order harmonic generation of some DG technologies, influence on power system stability, possible excessive voltage increase, short-circuit currents increase. While DG strengthens the distribution grid resulting in number of dips decreasing, the transmission network may weaken as described in Reference [28] and enforcement of distribution grid is necessary.

DG units have a significant role in European objectives, as presented in Reference [29], where the possibility of control and connection point is described with centralization and decentralization trends in European area. Moreover, Reference [29] gives The Smart Grid Architecture Model (SGAM) Framework that is to be respected in order to address interoperability in a way described in Reference [30]. Concise description of the SGAM Framework is given by Panda and Das [31], who address the still open questions of the SGAM and provide an insight of the Smart Grid environment in the year 2050.

3.1. SGAM Framework

SGAM Framework introduces six functional zones for specific purposes and functionalities, respecting user requirements and possibilities. First zone, Process, implies equipment that participates in energy conversion and transmission. Field zone, the second one, implies monitoring and protection equipment responsible for data acquisition. Aggregation level is in third zone, Station, in which substation automation supervises the first two layers. ADN management processed by this paper can be part of third zone, or, even better, part of fourth zone, called Operation.

Operation handles multiple Energy management systems (EMS), Distribution Management Systems (DMS), multiple microgrids, aggregation of various type RES into Virtual Power Plants (VPP) and electric vehicles (EV). The fifth zone, Enterprise, includes utilities, service providers, energy traders, customer relationships, asset management and procurement [29]. Market operations are in the sixth zone, Market, where energy trading and mass market occurs.

SGAM Framework also proposes the methodology which was used for this research to examine the validity and applicability of multiple CI methods in real-world power systems. The methodology within this research has seven crucial principles of validation:

1. Universality—The proposed solutions must be vendor-agnostic and without preferences in existing architectures and solutions. Neutral and scientifically objective perspective is mandatory when evaluating the possible solutions for future power networks.
2. Localization—Given the functional zones, the application of possible solution needs to be clearly stated for which zone it is intended. Systematic view of the whole

framework is needed to clearly understand the level in the topology in which any solution can be applied to.

3. Consistency—The use case must not violate the existing SGAM framework and in no case must single out one zone as dominant. Interdependence and reliable functioning with other zones must be respected.
4. Flexibility—Alternative designs and implementation of use cases, functionalities and services to any SGAM layer or zone need to be covered. Methods and algorithms observed in this paper can be part of any SGAM layer or zone, with the best starting point in the third or fourth zone.
5. Scalability—Once the proposed solution is part of one zone or a smaller part of the network, it must be able to successfully scale to other zones or to the whole network. Of course, the limiting factor here may be computational power, but this is not the subject of discussion in this paper and it can be solved by Cloud Computing [32].
6. Extensibility—Solutions need to adapt to evolutionary changes in the Smart Grid environment.
7. Interoperability—Interaction between multiple actors, applications and systems occurs via information exchange with defined protocols and data models. Chain of entities and connections define interfaces in which the consistency and interoperability become secured.

By combining Station and Operation zones with Communication, Information and Function layers in SGAM, it is possible to reach an enclosure in which there is an exceptional need to apply methods presented in this paper.

As mentioned above, in bullet 4, the good starting point for integration of CI-based methods for ADN management is Station or Operation zone in distribution system, DER and Customer environment. The bounding section for CI applications in ADN is given by Figure 1, where the application zone observed by this paper is shaded in red.

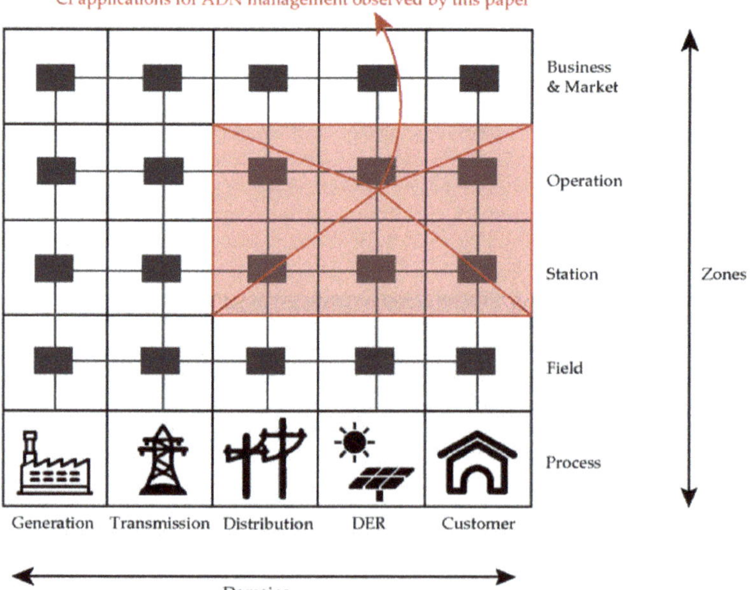

Figure 1. SGAM reference architecture with emphasis of CI applications observed by this paper.

Process and Field zones are most often oriented to smaller systems that can be explicitly mathematically described and that depend on a finite set of known parameters. Such systems are micro-grids of a local character or even small micro-grids at the end-point [33,34].

Large centralized generation connected to the transmission network is a separate category by itself, regardless of the zone of application. The laws of system management at these levels are completely different from the established methodologies of managing the distribution system, as can be understood from comprehensive overview given in Reference [35]. Interconnected power grids are multi-layered system where CI can be useful as a distributed optimization platform that acts as one part of the overall algorithm. This concept is explained in a little more detail later in this paper when one of the most applicable optimization methods is described.

Enterprise and Market zones can benefit from the solutions defined in the Station and Operation zones because such security of the right solutions will enable a different market presence and even develop new market possibilities, most often consequently with greater RES integration [36].

Considering that this paper is the result of the project that respects the SGAM architecture and according to the basic principles of the described methodology, it is possible to imagine the trend of developing inventive solutions for the Smart Grid environment. Such inventive solutions must by their nature react quickly enough to be taken into account and precisely enough for their application to come to life at all.

3.2. DG Impact in ADN

When considering CI methods for the application in system solutions complied with SGAM Framework, the properties of DG and the background mathematical models of each observed technology should be considered. The mathematical models of DG and the mathematical background of the power system represent a system of complex equations that are not easily solvable by conventional methods. A clear and comprehensive overview of the essential characteristics of different DG allocation methodologies types is given in Table 1.

Table 1. Comparison of main content of selected papers.

Reference	Optimization Method	Merits	Limitations
Acharya et al. [25]	Analytical approach based on mathematical expression	Simple analytical solution for determining size and position of DG	If DG is already installed there is a need for significant amount of power flow calculations
Gözel et al. [26]	Bus-injection to branch-current and branch-current to bus-voltage matrices.	One power-flow calculation, simple matrix algebra and faster performance	Does not take into account the interaction of nodes in the network when first DG is placed
Wang et al. [37]	Analytical approach	For radial feeders multiple DG can be taken into account; consideration of time varying DG	Only overhead lines with uniformly distributed loads are considered
Aman et al. [38]	Power Stability Index and Golden Section Search algorithm	Reduced computation time when compared to Golden Section Search Algorithm	Only radial networks are simulated and tested
Injeti et al. [27]	Simulated annealing	Comparison of meta-heuristic methods, including genetic algorithm, Particle Swarm algorithm and Loss Sensitivity Factor Simulated algorithm	Only radial networks are simulated and tested
Singh et al. [39]	GA	Pricing problem addressed instead of power losses	DG is considered as negative load, not simulated independently

Table 1. *Cont.*

Reference	Optimization Method	Merits	Limitations
Abou El-Ela et al. [40]	Modified GA	Consideration of voltage profile improvement, spinning reserve and line power flow reduction; exceptional mathematical overview of main DG impact indicators; tested on real Egyptian network	Time variability of generation and consumption not considered
Biswas et al. [41]	GA	Technical and economical optimization performed; visual representation of losses variation as DG location and size changes; essential problem formulation	Time variability of generation and consumption not considered
AlRashidi et al. [42]	PSO	Multiple DG planning and consideration in single run; fitness function development; backward/forward sweep approach; experimental design of custom PSO	Only time-independent loads and generation considered
Gomes-Gonzales [43]	PSO	Total power generation covers total demand and real power losses what makes foundation for island operation	DG units have fixed power factors; time-independent loads and DG considered
Moradi et al. [44]	GA + PSO	Hybrid method development; different algorithms benchmark	Time-independent loads and DG considered
Soares et al. [45]	Signaled Particle Swarm Optimization	Identification of active distribution network management; achieving a balance between production and consumption; hourly optimization; fast execution	Missing emphasis on island operation possibility and justification of that possibility
Saif et el. [46]	Double-layer Particle Swarm Optimization	Possibility of island operation and connected operation in which energy is sold to superior network; simulation-based optimization with considering stochastic behavior of distribution network; supply reliability considered	Sensitivity analysis of Particle Swarm Optimization model not emphasized
Kansal et al. [47]	PSO	Different types of distributed generation considered with Particle Swarm and analytical approach	Demonstration of Particle Swarm performance for various types of distributed generation is not clear

It is necessary to consider the fact that the data from the paper [12] published in 2012 have been changed today in the field of the installation cost. Furthermore, distributed generation allows reactive compensation control voltage level, can contribute to the regulation of frequency and can be used as spinning reserve during faults if the technology of production units permits [12,25–27,39,42,48,49].

Beside the technical advantages of the distributed generation, the authors of Reference [12] also provide an additional, indirect economic benefits expressed through the possibility of delaying investment in power system components and reductions of different power system management costs, while papers [12,48] refer reducing investment cost for the network equipment upgrade, reducing the operating costs of transmission and distribution systems, increasing the security of supply at critical loads, reducing the cost of mandatory reserves in the system, etc. The benefits of integrating DG units are reduction of losses, improvement of voltage profile and voltage stability, increased quality of electricity and efficiency of energy use, but direct current flows certainly have a negative impact on the reliability of the system [13].

Distributed generation in a whole with a consumption becomes a micro grid, while more micro grids make an active distribution network [50,51]. It can be concluded that DG change the technical properties of distribution network making this network an active distribution network—a precondition for development of Smart Grid where the distribution network has the ability to supply the consumers from distribution generation and upper voltage level network together while maintaining an optimal operational [50].

Distributed generation is the most useful when placed as close to consumers as possible to ensure the greatest impact on DG benefits. A terminal in the distribution network where DG has the greatest beneficial impact is called an optimal location. Selection of the optimal location is crucial for the planning of the distributed generation.

4. Optimal DG Planning and Scheduling

ADN Management is a real-time and operational planning problem when it is following functional criteria. Solutions for allocation problem described in the technical and scientific literature and papers can be used as scheduling problem if functionality, modeling constraints and performance rules are obliged. That interest requires the identification of papers which present solutions that enable development of a potentially new method for DG operational management planning in Smart Grid. Optimal ADN management with an increased share of RES-based and non-RES-based DG technologies is a modern challenge for the distribution system operators since most of the current distribution grids are not designed for the new, bidirectional operating conditions. ADN operational planning represents the necessary precondition to ensure the most favorable effects of distributed generation in Smart Grid environment [12,48,52]. Proper operations in Smart Grid manage power flows in a most efficient and reliable manner, but require integration of multiple technologies described in Reference [53].

4.1. Optimization Methods

Optimization methods that can be applicable to Smart Grid must comply with the requirements of real-time performance, consideration of technical requirements and actual production situation of various production capacities, while mitigating difficulties of weak convergence and local pinning [12,48]. The observed works in the literature distinguish five optimization-method groups [12,54]:

1. Analytical methods;
2. Heuristic and meta-heuristic methods;
3. Artificial Intelligence–based methods;
4. Evolutionary principles and biologically inspired methods;
5. Distinct methods for specific purposes.

Analytical methods require the definition of algebraic expressions basis of which can the optimization process be analyzed. Such approach will result with a well-developed mathematical model of the observed system. That model can be used in together with measurement data, but it is very demanding, almost impossible, to use it for more complex systems and challenges [12,26]. The advantage of the analytical approach is manifested through the usage of the model in conjunction with measured values and external input data. The common opinion in the literature is how optimization procedures can be divided into methods that result in precise solutions and methods that result in good-enough optimal Pareto solutions [55–58]. An overview and grouping of optimization methods is shown in Figure 2.

This paper observes best practices and methods that according to the authors can be used for real-world use cases. The usability is tried and proved by modeling and simulation in software engineering environment. This paper deals with the groups from Figure 2 and gives an overview of what has been tried, what has been repeated and what can be applied to real-world systems.

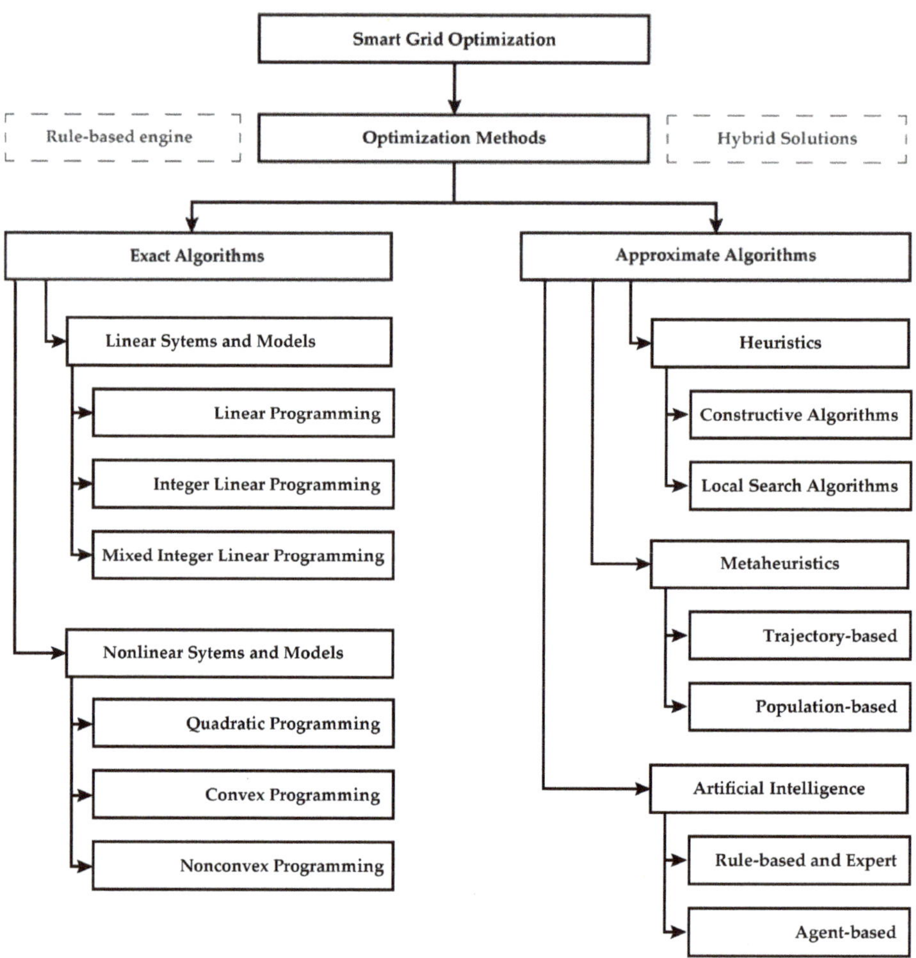

Figure 2. Optimization methods for Smart Grid applications.

A meta-heuristic approach is an iterative optimization process that can be assigned as a guidance for subordinate heuristics [59]. The heuristic process is beneficial for solving optimization problems in cases where a precise and accurate solution is not unique, when the set of acceptable solutions exist and when the precision of all solutions does not necessarily have to be absolute but good enough to be pragmatic enough [60]. Heuristic algorithms can be specifically developed to solve a particular problem or universally applicable with the common search criteria. In any case, heuristic algorithms can be used in a search for a specific set of possible solutions and finding out the best solution in the set. Meta-heuristic methods represent advanced procedures in which the heuristic process is further enhanced in every iteration of the solution search and most often apply universally applicable heuristic procedures [61]. The way of refining heuristic process with each iteration can be defined as a separate heuristic process and therefore such complex methods are called meta-heuristics. If the meta-heuristic optimization is complemented with the precise calculation results of individual solutions, it becomes possible to state that general strategies are specifically applicable, and the results are accurate and precise, thus using best of both worlds. For most meta-heuristic methods, it is necessary to properly represent the problem, procedures and operators within the legitimacy of performing the heuristics and to partially limit the scope of the solution. Metaheuristic methods that are significantly proven and applicable in solving various problems are local search, tabu search, simulated annealing, genetic algorithm (GA) and ant-colony algorithm [54,60–62].

Computational intelligence approach includes dedicated tools and intelligent procedures for solving certain types of problems that do not necessarily need to be optimization problems. Recognized examples of the methods based on the computational intelligence are fuzzy logic, artificial neural networks, rough sets theory, expert systems and probabilistic agents [63] and application of these methods is often necessary for the purpose of making decisions about individual values, sorting and numeric marking according to certain rules. The wider conception of computational intelligence comprises all heuristic and meta-heuristic optimization procedures and the boundary of the specific category where particular approach belongs to is often unclear.

Methods based on evolutionary principles are powerful optimization procedures most often used in optimization problems where is necessary to consider the fulfillment of two or more objectives. A common feature of optimization methods based on evolutionary principles is the existence of one or more decision-making procedures, the so-called evaluation according to fitness function [56]. Using iterative optimization process based on these methods it is possible to direct the search for solutions in each subsequent iteration to the area where the previous iteration achieved the best solution. Considering the above principle, it is possible to conclude that these methods are like meta-heuristic procedures, which additionally confirms the ambiguous boundary and complexity of the categorization of certain methods.

Hybrid methods combine two or more processes to a single optimization process. The most common hybrid methods are synthesis of the optimization process and computer intelligence decision-making procedures [60]. All optimization methods may encounter certain performance problems in finding the best global solution, low convergence, and long calculation time if they are not properly adapted to the observed problem [9].

4.2. Analytical Methods

Analytical methods are briefly addressed with the emphasis that their capabilities are limited and that advanced computer intelligence methods are better used in multi-objective optimization problems. In Reference [37], the authors show the possibility of analytical calculation of voltage drops and power losses instead of iterative load-flow calculation used to determine the node in radial or doubly fed feeder where DG will provide the lowest power losses. The analytical approach described in the mentioned paper observes two objectives; the location of the distributed generation to obtain the lowest power losses and to fulfil the target voltage values along the observed feeder. If the second objective is not satisfied, the location is changed near the original solution until both objectives are met. Power dispatched from DG authors limit with the injected current value that cannot exceed the power consumption from the DG location. The proposed method can be performed with the following load-distribution constraints: uniform load distribution in the feeder, centralized load distribution in the feeder load where the largest power consumers are in the middle of the feeder and loads are increasing towards the feeder. When solving the allocation problem in ring and interconnected distribution networks, authors use the admittance matrix and minimize the objective function (1) [37].

$$f_j = \sum_{i=1}^{j-1} R_{1i}(j)|S_{Li}|^2 + \sum_{i=j+1}^{N} R_{1i}(j)|S_{Li}|^2, j = 2, \ldots N \quad (1)$$

For the location of distributed generation in the node j, the aim is to find the lowest value of losses which is in the function of the equivalent resistance between the first and the i-th node. Additionally, the authors have investigated the possibility of renewable energy production volatility representation through a series of different data. Since the authors did not use iterative procedures there is no problem of convergence of the proposed method.

Achary et al. [25] are considering the GA usage in order to optimize the location of distributed generation but are abandoning the idea due to many iterative load-flow calculations which significantly affected the calculation time consumption needed for optimization. In their work, they use the theorem of the complex power that represents the basis for determination of the most sensitive node where distributed generation achieve the least loss in the system. The proposed solution according to the expression (2) will result in the optimal distributed generation dispatch for each node, while the contribution of each distributed generation to total system losses will be determine using theorem of the complex power. The node "I" with the least losses represents the optimal location of the distributed generation considering the consumed power "PDi" at that node.

$$P_{DGi} = P_{Di} + \frac{1}{\alpha_{ii}}\left[\beta_{ii}Q_i - \sum_{j=1,\, j\neq i}^{N} \left(\alpha_{ij}P_j - \beta_{ik}Q_j\right)\right] \tag{2}$$

Gözel et al. [26] developed a similar method to the one developed by the authors of Reference [14], with significant changes in the representation of the results in relation to the authors of Reference [25]. A significant contribution to their work stems from the definition of mutual influence of distributed generation node to other nodes in the distribution network. The results of their calculations are also based on power losses calculation as decisive and clear indication of distributed generation influence. In addition, the authors compare their results with the authors of Reference [25] and concluded that their method consumes less computing time.

Aman et al. [38] presented the Power Stability Index (PSI) method in order to find nodes in network which have the most favorable impact on voltage profile and grid losses when distributed generation is connected to them. The authors tested their method on a 69-bus test system and a significant contribution is shown defining the new voltage stability indicator for valuation of the nodes. PSI method is compared to Golden Section Search Algorithm (GSSA) and results are compared with strong similarities, but GSSA used more computation time when compared to PSI-based method given by expression (3), in which P_G represents active power of distributed generation, r_{ij} stands for real part of line impedance and V_i is the real voltage of the i-th bus with the voltage angle, δ.

$$\text{PSI} = \frac{4r_{ij}(P_L - P_G)}{[|V_i|Cos(\theta - \delta]^2} \leq 1 \tag{3}$$

4.3. Computational Intelligence–Based Methods

Significant development of CI and numerous papers have pointed the applicability of selected metaheuristic methods used as higher order procedures to determine sufficiently good solutions without the necessary knowledge of the entire mathematical model or the values of all variables. Solutions derived from metaheuristic approaches are result of search for global best solutions in the predesignated and finite search space [64,65]. If the search space consists of many possible solutions and solution variants, metaheuristic methods often repeat some of the properties and re-evaluate already visited space. That feature often results in more precise and refined solution. Classification of metaheuristic methods is given by Figure 3 which presents most viable algorithms, grouped by type, governance rules, modeling rules and the way they determine the best solution.

Computational intelligence implies hybrid approaches created by combining several optimization methods and is characterized by successful application for continuous values, the ability to self-evaluate and change the way of execution, stochastic approach, parallelism in execution and the ability to generate approximate or Pareto solutions [66]. Procedures and methods contained in computer intelligence often base their principles of operation on biological principles or natural processes [67] and are applied in solving problems for which there are no effective or specialized procedures [68]. The lack of specific procedures for specific problems may be caused by complexity that does not allow effective modeling or the inability to explain or model certain problems and all observed factors [69]. In power engineering, computer intelligence can be used to solve many optimization challenges, calculate optimal power flows [70–72], system modeling and monitoring of ADNs using fuzzy logic systems with evolutionary algorithms and artificial neural networks, while some procedures and algorithms can be used in data and event analysis and diagnostics of ADNs using qualitative reasoning, planning methods and hybrid procedures [73]. In addition, computer intelligence is successfully used to solve problems of designing electromechanical oscillation stabilizers, determining the causes and sorting of faults in the transmission network, reliability assessment, consumption forecasting, power system protection coordination, electricity quality assessment, economic supply of electricity, reactive power optimization and determination of optimal power flows—basically everywhere where iterative processes need faster convergence [74].

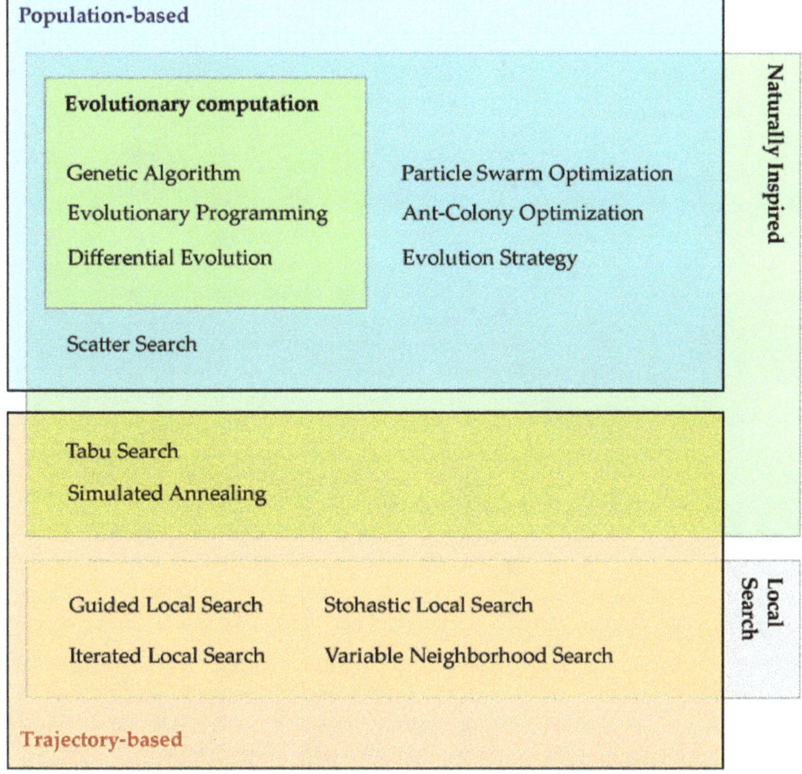

Figure 3. Metaheuristic methods classification.

An extensive review of optimization procedures used in mathematical modeling in 360 scientific papers was provided by Theo et al. [7] outlining the advantages and disadvantages of each approach. For a GA, the authors outline suitability for solving problems that may have more favorable solutions, it is generally easy to integrate into an existing simulation framework, it is tolerant to the objective functions with chaotic attributes, and it is suitable for topological and categorical variable optimizations. However, as a lack of a GA authors outline the convergence to the local best solution instead of the global best solution, long-term convergence and complex approach of determining the criteria for the optimization process termination. When reviewing the Particle Swarm intelligence algorithm, the authors outline the advantages of fast execution, flexibility and openness to the other soft computing procedures but warn that the algorithm requires definition of coordinate motion system and a proper number of particles otherwise can result in local best solutions.

A comprehensive overview of optimization methods that can contribute to the more efficient integration of DG into the power system are given by Colmenar-Santos et al. [75] in their paper from 2016. The authors of Reference [75] state that the problem of the multi-objective optimization of various technology DG integration into a fully developed active network has not yet been completely solved and outline the idea of the development of a robust distribution management system called AMN (Active Management Network) whose role is a real-time operational management of DG and other control devices in distribution networks. Analyzing the optimization methods and grouping seen in the literature, the authors of this paper represent the division of optimization approaches into three groups:

1. Conventional approaches.
2. Approaches based on the artificial intelligence.
3. Hybrid approaches based on multiple methods of artificial intelligence.

According to Reference [75], conventional approaches are analytical methods, power flow calculations, non-linear programming method and rule 2/3; approaches based on artificial or computer intelligence consider evolutionary algorithms, algorithm of simulated annealing, differential evolution algorithm, Particle Swarm algorithm, fuzzy logic systems, ant-colony algorithm, tabu search algorithm, artificial bee colony algorithm, firefly algorithm while hybrid approaches include methods of unification of the GA and fuzzy logic systems, GA and tabu search algorithm, GA and Particle Swarm Optimization (PSO) algorithm, GA and power flow calculation, PSO algorithm and power flow calculation, tabu search algorithm and fuzzy logic systems.

After reviewing all of the methods, the authors [75] state that solutions based on the swarm intelligence algorithm are complex in development if reliable global solutions are sought while fuzzy logic systems and hybrid systems are not sufficiently represented in the literature. The authors outline that premature convergence is extremely emphasized for the methods based on the evolutionary principles if solution approach is not detail enough, while other significant methods, such as the simulated annealing algorithm and ant-colony algorithm, have long execution time which excludes them for possible application in short-term planning.

The largest number of scientific papers where meta-heuristic optimization methods are applied for allocation of distributed generation use the GA method or the PSO method, and there are also papers that use a hybrid method as a compound of these two procedures as well. Singh and Goswami [39] use GA determined by the objective function (4) for solving specially designed multi-objective optimization problem, where the impact of the distributed generation integration on active power market price λ is considered beside the power losses reduction and voltage conditions improvement. The authors used the knowledge presented in Reference [76] when assuming distributed generation influence on the electricity price $C_i(DG)$, depending on consumer active power P_{Di} and distributed generation active power P_{DGi}.

$$P_{eleect}^{DG} = \sum_{i=1}^{n} [\{C_i^a(DG) \times (P_{Di} - P_{DGi}) \times \Delta t + C_i^r(DG) \times (Q_{Di} - Q_{DGi}) \times \Delta t\} + \{C(DG) \times P_{DGi} \times \Delta t\} + \lambda \times P_L(DG) \times \Delta t] \quad (4)$$

The presented considerations of the authors of Reference [39] successfully confirm the simulations on the radial distribution network model, where they prove the usefulness of the proposed method for the allocation of one or more distributed generation in radial distribution networks.

4.3.1. GA-Based Methods

GA is a space search heuristic that is inspired by biological process of natural selection in which the most potent individuals of one generation give birth to individuals of the new, improved, generation. Genetic sequences determine the possibilities of the individuals within the population. The individuals are rated based on fitness function and the most fit ones reproduce to design a novel individual with better set of genes. If modeled correctly, GA can bring real adaptive computer programs. Main genetic operators are crossover and mutation, with all its advantages and disadvantages and difficulties in coding physical and mathematical models from the power industry domain as genes [77]. Binary encoding can be used in discrete search areas, while for continuous values a real-value encoding or tree encoding scheme needs to be used [77,78]. The process by which binary GA operates, as described in Reference [79] is presented by Figure 4. The process starts with objective function formulation, as with any method, but after the initial population the GA operators take place. In this example, each individual is encoded with n number of genes and each gene consist of m number of bits, as given by Figure 4. The fitness evaluation is usually in power engineering done my means of co-simulation and calculations. GA operators are mandatory as they ensure that the algorithm observes the whole search space.

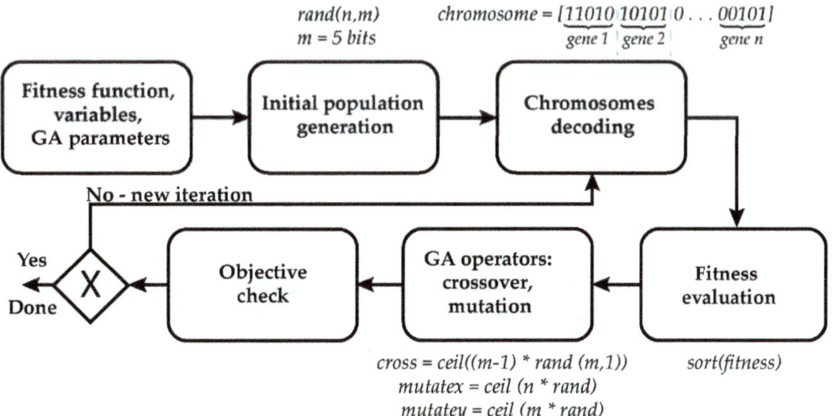

Figure 4. Binary GA general process diagram.

A fundamental feature of GA is competition among individuals and the principle of elitism. Elitism implies the survival of only those individuals who have a high fitness score, and all other individuals are rejected. When observing multiple objectives simultaneously, it is necessary to pair individuals that have an elitist grade according to one of the criteria, in order to obtain an offspring that can have elitist grades according to all criteria. In some cases, most often when multiple solutions have equal value, not numerically, but value to the observer, selection methods such as Tournament Selection and Rank Selection are used in which individuals compete among each other in random manner to prove the unquestionability of the best individual.

Injeti et al. [27] used simulated annealing method for optimization and compared results with results obtained with GA and PSO method. In Reference [27], the authors clearly stated constraints that should be considered when using advanced optimization methods of computational intelligence such as simulation limitations. Successful implementation of one method of computer intelligence and review of other methods used for allocation of distributed generation are clear indicators of the future research, as shown in the more recent paper of the same authors [80].

$$P_{T,Loss} = \sum_{i=0}^{n-1} P_{Loss}(i, i+1) \tag{5}$$

$$P_{Loss}(i, i+1) = R_{i,i+1} \cdot \frac{P_i'^2 + Q_i'^2}{|V_i|^2} \tag{6}$$

In Equations (5) and (6) $P_{T,Loss}$ stands for total feeder losses and P_{Loss} represents line losses, the main minimization objectives considered in Reference [27].

Abou El-Ela et al. [40] used a method based on the GA for DG allocation and examined the proposed method by linear programming. The same authors concluded that there is no significant deviation in the obtained results, thus confirming the usefulness of computational intelligence–based methods. The authors of Reference [40] define clear indicators with or without DG for voltage profile improvement—VPI according to (7); spinning reserve increasing—SRI according to (8); power flow reduction—PFR according to (9); and total line loss reduction—LLR according to (10) [40]. The developed-algorithm authors successfully tested on a model of a real distribution network in Egypt.

$$Max\ \text{VPI}\% = \frac{VP_{w/DG} - VP_{wo/DG}}{VP_{wo/DG}} \times 100 \tag{7}$$

$$Max\ \text{SRI}\% = \frac{SR_{w/DG} - SR_{wo/DG}}{SR_{wo/DG}} \times 100 \tag{8}$$

$$Max\ \text{PFR}\% = \frac{PF_{k,wo/DG} - PF_{k,w/DG}}{PF_{k,wo/DG}} \times 100 \tag{9}$$

$$Max\ \text{LLR}\% = \frac{LL_{wo/DG} - LL_{w/DG}}{LL_{wo/DG}} \times 100 \tag{10}$$

The benefits of DG are present with index MBDG—maximal composite benefits of DG—according to the authors of Reference [40]. For the limitations of the computer intelligence optimization process, the authors state the limitations are the maximum number of distributed generation units so in their paper authors analyze the distribution of one distributed generation in the distribution network and conclude that the nominal power of the power plant significantly affects the indicators determined by expressions (7)–(10) unlike the location—node—of the power plant. The authors present the energy balance of the observed system as a limitation of the optimization procedure and determine the condition according to which the power of total production in the observed network must be equal to the sum of all time-constant consumers and power losses. Similar authors use GA to allocate the remote measurement and monitoring units in the Smart Grid environment Reference [81], where they deeply explain the parameters of GA-based optimization process.

Biswas et al. [41] used genetic algorithm for the optimal DG allocation in order to reduce voltage sags in a real complex distribution network. Presented problems are perceived by the authors through the real power losses (RPL) reduction objective function (11) that takes into account quotient of product of voltage angle difference cosine and line resistance with voltage level A_{ij}, and quotient of product of voltage angle difference sinus and line resistance with voltage level B_{ij}; objective function for decrease of customers that can be affected by the voltage sags in observed time period (NF) (12) with the total load distributed, S_{DIST}, and the load distributed for the i-th fault, L_{DIST_i}; the objective function of DG integration costs reduction (13), which is determined by financial units per kilowatt of installed DG power (KC) and observed as the objective of determining the least power of DG which will meet the previous two objectives [41]. The constraints of the proposed algorithm that the authors have defined, similarly to previous authors, are load-flow constraints through the specific branches, voltage constraints and limitations of the number of DG units.

$$\text{RPL} = \sum_{i=1}^{N_b} \sum_{j=1}^{N_b} A_{ij}(P_i P_j + Q_i Q_j) + B_{ij}(Q_i P_j - P_i Q_j) \tag{11}$$

$$Min\ S_{DIST} = \sum_{i=1}^{N_F} L_{DIST_i} \tag{12}$$

$$Min\ C_{DG} = K_C \sum_{i=1}^{N_F} P_{DG_i} \tag{13}$$

The specifics of the proposed method based on the genetic algorithm is emphasized in the integrated use of power flow calculations when evaluating the benefit of the best solutions in each iteration. After evaluating the benefit of individual solutions based on the power flow calculation, an additional algorithm of voltage failure analysis is performed, which evaluates the impact of the proposed solutions on the number of time-constant consumers covered by voltage failures. This approach is innovative because it considers multiple objectives taking place as several separate optimization procedures. The way in which the authors graphically present the results of several iterations with a three-dimensional surface is seen later, in other works.

The limitations of methods based on a genetic algorithm for solving more complex optimization problems are explained in the paper by Yang et al. [18] in which a hybrid method of two genetic algorithms is used with the aim of determining the maximum power of distributed generation, taking into account the voltage and technical limitations of the elements of the distribution network. The first genetic algorithm, determined by the objective function (14), determines the minimum power of distributed generation limited by the short-circuit power on the primary side of the competent transformer, voltage level and technical characteristics of the equipment in the system. The second genetic algorithm, according to the function of the goal (15), determines the power of distributed production that will meet the requirements of consumers of the observed derivative, while respecting the solutions of the first genetic algorithm.

$$Min\ f_{1,i} = w_1 \left(\sqrt{\sum_{j=1}^{k} \left(P_i \left(S_{MVA_j} \right) \right)^2} \right)^{-1} + w_2 \left(\sqrt{\sum_{j=1}^{k} \left(P_i \left(S_{MVA_j} \right) \right)^2} \right) \tag{14}$$

$$Min\ f_{2,i} = w_1 \left(P_i \left(S_{MVA_j} \right) \right)^{-1} + w_2 P_i \left(S_{MVA_j} \right) \tag{15}$$

Using this hybrid approach and using two genetic algorithms, the authors determined the possible power intervals of distributed generation depending on the short-circuit power on the busbars of the parent network. Moreover, same authors showed that in some specific cases the usual meta-heuristic methods may give false solutions.

Methods based on genetic algorithms that solve optimization problems of reactive energy generation with the objective of reducing losses in the observed systems are particularly presented in papers [82,83] and the paper of the author López-Lezama [84] where the optimal distribution of the DG is determined according to the electricity price criterion. In Reference [84], the advantage DG is presented through the possibility of achieving a higher generated power price from the market price due to the fact that DG is near to consumers and that feature results in fewer power losses. The same authors emphasize that repeated optimization procedures with the random initial settings resulted in almost identical solutions, thus confirming the usefulness of the proposed method.

4.3.2. PSO-Based Approaches

PSO is a population-based, biologically inspired method with a limited search space, that can repeatedly result in usable optimization results. The intrinsic feature of the method is the re-evaluation of an already past search space. Each possible solution in PSO is called particle and it is described by position, mass and velocity. Particle has its own memory of personal best solution and communicates with the group of particles to examine group's best solutions. PSO initialization implies random particles at random places and with random parameters that are updated with each iteration. Particles follow their own fitness criteria, called personal best or pbest, but comply with the group's fitness goals, called group best or gbest. Depending on the fitness value, the particle follows the group, or vice versa, as described by Figure 5.

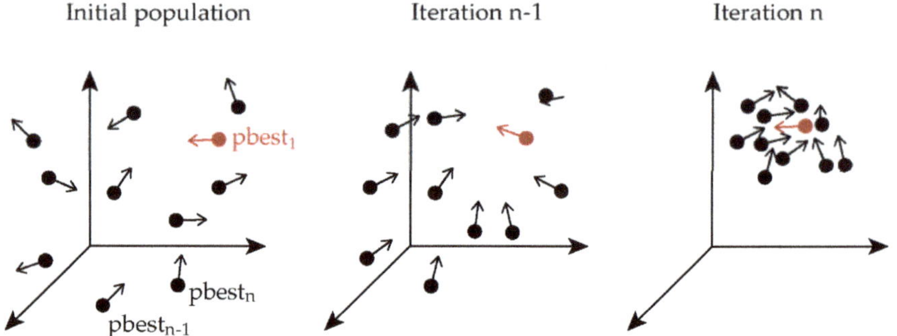

Figure 5. Particle following in PSO.

There are cases where particles take the entire population in the wrong direction or where a population converges toward a solution that is not globally best. This most often occurs when the number of particles and parameterization in the population does not correspond to the observed problem. When observing a challenge in a Smart Grid environment, it is possible to easily solve wrong convergence by determining the number and parameters of particles according to the observed electric power system. This means that the number of particles in each generation should at least correspond to the number of observed cases, and their random parameters should be limited to the uniqueness of the search space.

Figure 6 gives a general overview of the PSO method process. At first it seems that the PSO has more steps than GA, but given that PSO evolves around single swarm representation, it shows better performance in most cases. Random swarm generation should in most cases consist of at least same number of particles as there are discrete possible solutions—in case of ADN and DG that number represents number of buses. Each particle has its own velocity, which is encoded as power generation. When a swarm identifies a spot, or a bus in distribution network, in which the DG is most necessary, that becomes a new swarm orientation. Similarly, when a particle finds a most appropriate power is more that becomes the global information. Particles are then updated with the new information and the search space gets explored by the complete swarm.

In Section 3.1 of this paper, the possibility of distributed optimization platform was mentioned for large power systems. Each CI-based solution in an interconnected multi-layer power grid can be represented as a single particle of an overall PSO-based power system management solution. Of course, at the time of writing this paper, this idea is only a far theoretical consideration, but such were once the ideas and concepts discussed in this paper, and today they are the basis for the development of advanced software solutions. It will certainly be interesting to witness the development of modern power engineering and the growing representation of CI methods in modern solutions.

A novel approach for solving problem of multiple DG units distribution suggest the authors Alrashidi et al. [42] with the PSO algorithm as a basis of their method. The proposed method is confirmed on the model of the distribution network with several feeders and ten nodes where the optimal distribution of DG units have been determined through the multiple scenarios with the power losses reduction objective according to (16).

$$\min P_{Losses} = \frac{1}{2}\sum_{i=1}^{NB}\sum_{j=1}^{NB}\Re\{y_{ij}\}\left[|V_i|^2 + |V_j|^2 - 2|V_i||V_j|cos\delta_{ij}\right] \quad (16)$$

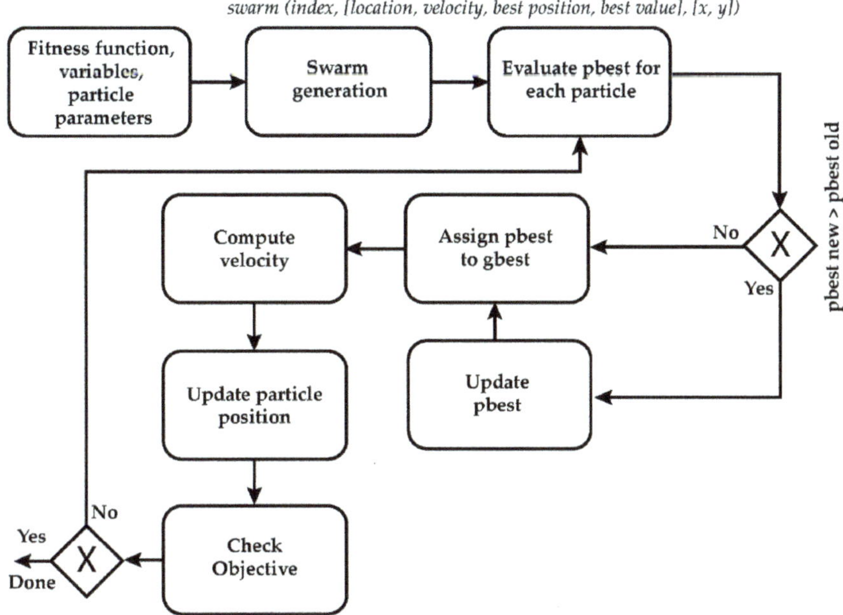

Figure 6. General PSO process diagram.

The constraints of the algorithm of the authors of Reference [42] define through the voltage constraints, load-flow limitations, size of the transformers and maximum permitted number of DG units. The authors especially emphasize that their method requires consideration of only one DG per node and only time-invariant production and invariant consumers can be. This paper, together with Reference [85], represents a framework of the development of a new method based on the Particle Swarm intelligence algorithm that will solve the problem of multiple DG multiple per node by considering the time-varying feature of loads and RES-based DG units. El-Zonkoly [85] defines the algebraic indicators for the objective function used for multi-objective optimization problem solution. Such indicators in the observed system are the influence of DG on active (ILP) and reactive (ILQ) power, the influence of DG on the voltage conditions, the possibility of DG dispatch in dependence of DG location and consumer location (IC), the influence of DG on short-circuit currents (ISC). For each defined indicator, the author assigns the weight factor σ whose variation opens the possibility to orientate the set of solutions to system management solutions or to system development adapted solutions [86–88]. Multi-objective function (MOF) thus becomes the sum of the indicator's multiplications and the corresponding weight factors according to expression (17).

$$\text{MOF} = (\sigma_1 \cdot \text{ILP} + \sigma_2 \cdot \text{ILQ} + \sigma_3 \cdot \text{IC} + \sigma_4 \cdot \text{IVD} + \sigma_5 \cdot \text{ISC}) + \text{MVA}_{sys(pu)} \tag{17}$$

Such an approach is common in the scientific literature and represents the combination of several individual functions into one objective function. In that way, a modeled optimization problem can be considered both single-objective and multi-objective optimization. Actual multi-objective optimization implies a vector of functions and the existence of opposing solutions.

In the second paper of the same author [88], the problem of the system load peak following is solved by the optimal DG allocation using the method based on the artificial bee colony algorithm, a subset of PSO algorithm. The proposed algorithm addresses the DG distribution by considering the criterion that the most demanding consumers are designated as primary for the DG integration. In this paper, the author considers three DG units in a system with 45 nodes or three DG units in a 33-node system while the installed DG capacity does not exceed 30% of total system load. A significant advantage of this paper is the consideration of two types of DG technology, one time-varying and unmanageable and other manageable DG. Moreover, the author anticipates the use of electricity storage system to reduce the required commitment of manageable DG.

Authors Gomez-Gonzalez et al. [43] represents an integration of load-flow calculation with an algorithm based on Particle Swarm intelligence to determine the optimal distribution of DG units, as presented by Figure 7. The authors of Reference [43] used the frog-jumping algorithm (18), which belongs to the group of PSO algorithms. Because of its specificities during particle coding, the algorithm is better adapted for the optimization of discrete values.

$$P_{\omega}^{t} = p_{\omega,max} - \frac{(t-1)\cdot(p_{\omega,max} - p_{\omega,min})}{(t_{max}-1)} t = 1, 2, \ldots, t_{max} \tag{18}$$

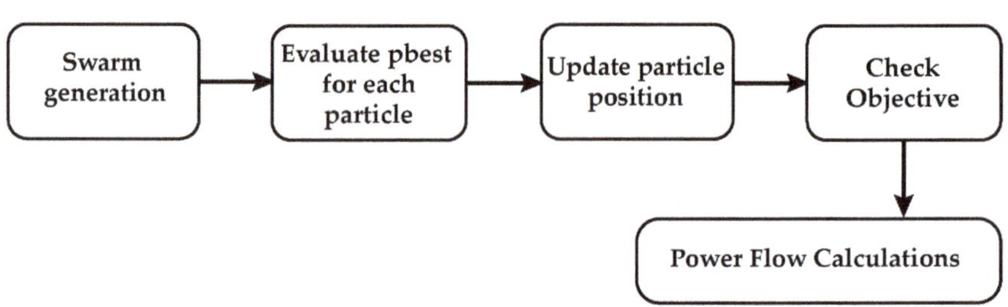

Figure 7. Simplified illustration of PSO and Power Flow co-simulation by Gomez-Gonzalez et al.

Significant contribution of this paper is a determination of an optimal number of DG units which must result in the ability of the observed system self-sufficiency. This emphasizes the value of this paper compared to previous papers in which the location and the capacity of only one DG unit were considered.

Moradi and Abedini [44] produced a new method based on a synergy of GA and PSO. The authors propose the genetic algorithm for optimal DG location determination and Particle Swarm Optimization technique for adequate DG capacity. This paper wisely used GA's possibility to perform better for integer-based, binary encoded, problems and continuous orientation towards the best solutions specific for the PSO algorithm.

Interestingly, the authors examined three separate cases, one with PSO only, one with GA only, and one with the proposed GA/PSO combination. According to the results presented in their work, the highest value of the objective function was achieved by a separate GA, while PSO and GA/PSO gave equal values of the objective function. However, the variance of the solution in PSO is many times less than in GA, while in the combination of GA/PSO variances in the objective function are negligible compared to any other algorithm.

Simplified flowchart of the presented hybrid method is given by Figure 8 by which the statement how the discrete values of the buses in the power system is a mitigatory circumstance is confirmed. The authors Moradi and Abedini used GA for DG location observations; since the DG location can be only on a system bus, it cannot be somewhere in between, and PSO is used for continuous optimization of DG power once the bus is identified.

The fulfillment of the technical conditions is defined by the algorithm constraints that are integrated into a common function according to (19). The function f_1 represents the objective of real power losses reduction, the function f_2 represents the objective of the voltage profile improvement while the function f_3 is the radial feeder voltage stability index taken from the literature. Coefficients k_1, k_2, β_1 and β_2 represent the penalizing coefficients of the corresponding expressions. The presented objective function considers the values of the voltage (V_{ni}) and the apparent power (S_{ni}) on the observed radial feeder busbars.

$$f = Min\left((f_1 + k_1 f_2 + k_2 f_3) + \beta_1 \sum_{i \in N_{DG}} \left[max(V_{ni} - V_{ni}^{max}, 0) + Max\left(V_{ni}^{min} - V_{ni}, 0\right)\right] + \beta_2 \sum_{i \in N} max(|S_{ni}| - |S_{ni}^{max}|, 0)\right) \quad (19)$$

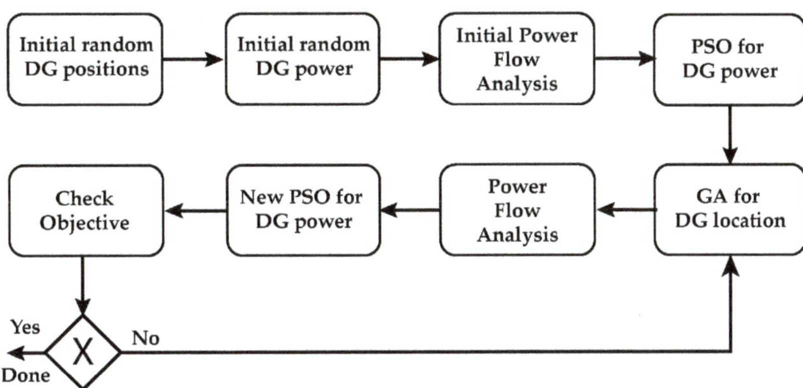

Figure 8. Simplified illustration of PSO and GA hybrid method by Moradi and Abedini.

According to the authors, the proposed method shows a significant increase in the accuracy and repeatability of the results, but the authors conclude that more time is needed for the method implementation compared with using one of the methods separately.

An innovative approach to optimization is offered by Saif et al. [46] defining a simulation-optimization process where the technical validity of the optimization is evaluated by a simulation check instead of the optimization process limited by equality or inequality. The developed method, which authors call dual layer simulation optimization because of its simultaneous performance in the MATLAB and the GAMS programming tool, uses the load-flow calculation for all simulation cases and Particle Swarm method to determine the distribution of different types of unmanageable DG with objective to increase the reliability of the system supply and to reduce operating and investment costs. The developed method is successfully implemented on the part of the real power system. Surprisingly, among the optimal DG distribution solutions solar power plants and electricity storage systems are not included, but only a wind power plants which authors attribute to the availability of primary energy in the observed area. The conclusion of their work is to meet the energy needs of the observed network with multiple radial feeders using unmanageable DG thus creating the precondition for the island operation of a future Smart Grid.

Two papers by Aman et al. [49,89] describe the use of the PSO algorithm with the objective of scheduling multiple DG units. The authors introduce new criteria for describing the impact of a DG unit on the observed network—the criteria of possible increase of load and index of feeder stability. By additional checks and comparisons of the proposed method with the analytical method and the search algorithm, the authors confirm the correctness of the use of the PSO algorithm to solve the DG scheduling problem with losses reduction objective thus achieving the most useful impact of DG on the observed network.

Comparison of optimization methods developed on the basis of the PSO algorithm and the analytical approach were presented in the paper of Kansal et al. [47] which addresses the problem of distribution of different DG technologies with objective of reducing losses according to (20).

$$Min\ P_L = \sum_{i=1}^{N}\sum_{j=1}^{N}[\alpha_{ij}(P_iP_j + Q_iQ_j) + \beta_{ij}(Q_iP_j - P_iQ_j)] \tag{20}$$

Authors define four DG types, considering the possibility of generating active and reactive power and suggest integration of one of them with respect to the results of the simulation. The results obtained with different optimization approaches are identical in the analysis of the optimal DG location and vary up to 5% for the optimal DG power dispatch.

In the leading scientific bases exist several scientific papers that use biologically inspired optimization algorithms. Niknam et al. [90–92] published several papers where honey-bee mating algorithm is used to solve the multi-objective problem of DG distribution in the distribution network [90] or to solve the problem of change of distribution network topology [91,92]. Equations from Reference [90] are given by (21)–(24):

$$F_1(X) = \min\sum_{i=1}^{N_{fc}} C_{fc}\left(P_{fc}\right) + \sum_{i=1}^{N_{wind}} C_{wind}(P_{wind}) + \sum_{i=1}^{N_{pv}} C_{pv}(P_{pv}) + Cost_{sub} \tag{21}$$

$$F_2(X) = min\sum_{i=1}^{N_{bus}}\frac{|V_{Rating} - V_i|}{V_{Rating}} \tag{22}$$

$$F_3(X) = min\sum_{t=1}^{N_d}\sum_{i=1}^{N_{br}}\left(R_i \times |I_i|^2 \times \Delta t\right) \tag{23}$$

$$F_4(X) = \min E_{t_{fc}} + E_{t_{wind}} + E_{t_{pv}} \tag{24}$$

In the papers authored by Niknam, according to the principle of selection of the queen bee among the bees, the best candidate for the survival of a population in the power engineering represents the DG or network topology.

The usefulness and efficacy of the PSO algorithm is further clarified in References [70,93–99], where this algorithm has been used to solve optimization problems or prediction problems and algorithms of evolution strategy have been compared with this algorithm.

Authors Kumari et al. [70] compared the GA, the improved GA and PSO algorithm for optimizing static VAR compensators in the network with the objective of reducing losses and achieving optimum load flows in the network. The improved GA suggested by authors is highlighted by the implementation of five additional gene crossing operators, in addition to the three commonly used ones, and according to their results, it provides the best optimization results.

Authors Yin et al. [93] used the PSO algorithm to optimize the allocation of computer processes to distributed processors, and this work, although in the field of computer science, thoroughly clarifies the application of the used algorithm. With the hybrid method which arose from the combining of the PSO algorithm and fuzzy logic systems authors Zhang et al. [94] solve the problem optimal reactive power flows to maintain voltage in transmission networks. The best solutions of each iteration of the Particle Swarm intelligence algorithm in this paper have been used as a measure of the feature parameters for the next iteration.

The adjusted PSO algorithm that addresses the problem of optimal DG allocation with the objective of losses reduction in the observed system is presented by Ashari et al. [95] while the usual Particle Swarm intelligence algorithm approach to solve the same problem is used by Bhumkittipich et al. [96] for the similar problem. A complete review of all derivatives of the PSO algorithm that have been known to authors Zhang et al. [97] and which can result in a number of best solutions, points to the usefulness of the PSO algorithm to solve many optimization problems. The apparent similarity of the algorithms in the papers of various authors [44,98] suggests that it is possible to achieve somewhat different approach and custom algorithm with exceptionally small variations in the principles of performance. That is also warned by authors [59] who advocate for a different development process of meta-heuristic methods and more significant changes.

A hybrid approach for addressing the DG distribution with the objective of total losses reduction and maintenance of a complex distribution network voltage profile, based on a unified GA and artificial neural networks, along with the GA and power flow calculation, is presented in Reference [100]. In this paper, the genetic algorithm determines the optimal location of one DG in a distribution network while the DG impact on the total losses is determined by the artificial neural network integrated into the GA. Described approach has not resulted in valid solutions when addressing the DG distribution problem in ring topology of distribution networks and a new hybrid approach has been developed by combining the genetic algorithm and the power flow calculation. In the same paper is presented a hybrid algorithm developed on the unification of two artificial neural networks with operating principles based on the search algorithms, but the authors point to the disadvantages of such an approach.

The need for active distribution network management is mentioned for the first time in the scientific paper of authors Soares et al. [45], who recognized the need for answering to consumer demands and the periodicity of some energy resources. The authors presented a special form of signaled PSO algorithm that enabled particles to change the speed parameter during the optimization process to gain the most feasible combination of wind, photovoltaic, fuel cell and energy storage. Change of speed was defined by given rules and observed with demand response and optimal storage charge/discharge. The proposed method was successfully tested on a model of Portugal's power system. Use of such precisely designed meta-heuristic optimization methods in market environment and real-life conditions was proven, and the direction of future scientific research was given in the form of one-day division in 24 independent simulations, which later became the paradigm of all future research. Although the proposed method did not provide the most economical result, but the second of six different methods, the most efficient method mixed-integer nonlinear programming (MINLP) is not suitable for larger systems and is not fully usable in production [45]. The authors presented with great quality a mathematical model of contemporary challenges in the power system, given by expression (25).

$$\begin{aligned}\sum_{W=1}^{N_W} P_{Wind(w,t)} + \sum_{PV=1}^{N_{PV}} P_{Photovoltaic(PV,t)} \\ + \sum_{FC=1}^{N_{FC}} P_{FuelCell(FC,t)} \\ + \sum_{S=1}^{N_S} P_{StorageDischarge(S,t)} + \sum_{LC=1}^{N_{LC}} P_{LoadCurtailment(LC,t)} = \sum_{L=1}^{N_L} P_{Load(LC,t)} + \sum_{S=1}^{N_S} P_{StorageCharge(S,t)}\ ;\ \forall t \in \{1, \ldots, T\}\end{aligned} \tag{25}$$

Expression (21) lacks the observability of power losses in power equipment, such as cables and transformers, that can be described as $\sum_{PL=1}^{N_{PL}} P_{PowerLosses(PL,t)}$ and added to the right side of the equation for the part of the Smart Grid to be self-sufficient. Observed units PV, FC, S, LC and L present appropriate technologies used in period t.

Presented scientific papers are clear indicator of the possibility for the development of robust intelligent solutions that can solve a certain set of complex problems in the power engineering, but without needed criticism, approaches based on soft computing methods should not be accepted as general solutions for absolutely all cases [101]. The authors also gave a mathematical model of technical limitations of the power system as inequalities that consider the specifics of individual technologies.

Reliability of distributed generation supply was introduced by Saif et el. [46] who presented double-layer simulation optimization, using load flow for all simulations and Particle Swarm Optimization in order to increase reliability of observed system. The authors optimized the given problem by aiming for operating and investment cost reduction. Developed method was successfully tested on a British power system model and the results were compared to other optimization methods. The authors of Reference [46] identified conditions for isolated distribution network operation. Aman et al. [38,49] used standard Particle Swarm Optimization algorithm, but introduced new criteria for describing the impact of distributed generation such as increasing the load capacity of observed system and the stability of network feeder.

The presented papers clearly point to the logical conclusion that, the once advanced and sometimes obscure methods known only to the scientific community, with today's computing power, can be used effectively in real production and be integrated into tools to optimize the distribution network in real time. Detailed mathematical models with metaheuristic methods supported by iterative calculations with thoroughly examined features and shortcomings of individual methods promise a new advanced Smart Grid in the future.

4.3.3. GA and PSO Comparison

GA and PSO share many common features, such as random start, fitness values, population update based on randomness and necessity for fine-tuning in order to find global optimum [57]. However, PSO does not have crossover and mutation operators and particles have memory and velocity. Main difference between two methods is that GA comes with new solutions with struggle among individuals, while PSO nurtures social interaction between particles.

Social interaction in PSO defines natural leaders that spread information to others, while in GA chromosomes share information with each other and guide the whole group towards distinct area, leaving only elite individuals in each iteration. Moreover, population in GA gets smaller with each iteration since some of the individuals become removed, while in PSO population is constant and particles always have purpose in fine tuning the possible solution.

PSO is derivative-free, robust and flexible method, but prone to premature convergence if tuning parameters are not correct. However, particle parameters can be enhanced with solid mathematical equations that improve analysis and provide realistic convergence conditions. Self-organization of PSO with bottom-up approach integrated in the method prove wide applicability in many scientific areas.

Global perspective of each particle in PSO improves clustering efficiency and enables PSO algorithm to test multiple areas of the search space at the same time. Instead of competition-based GA, PSO values the cooperation of particles and the exchange of values according to objective criteria.

For greater overview and comparison Table 1 gives a perspective on solved challenges in ADN and SG environments. In accordance with the Chapter I—Introduction of this paper, Table 1 clearly gives a standpoint how the Distribution Network Optimization includes many topics and issues identified in the modern distribution networks, such as feeder reconfiguration, island operation, pricing and energy market, voltage profile improvements, etc.

5. Optimal Smart Grid Management

Intuitive thinking about the purposefulness of presented papers and research is confirmed by studying features in the field of research in current European and world research activities. Since 2009, the Working Group for Advanced Power Grids of the European Commission Smart Grids Task Force has been continuously issuing guidelines for the development of modern power systems [102].

Advanced power systems imply a change in paradigm by enabling DG, but also the application of technologies for monitoring, operational management, regulation of production and consumption using information technologies that are not the subject of this scientific research. The Joint Research Center for Smart Electricity Systems and Interoperability leads regulatory, technical and economic research at the level of the European Union. According to the relationship matrix [103] of the Joint Research Center, the countries most involved in research into advanced power systems are Spain, France, Italy, Germany, the

United Kingdom, Belgium and the Netherlands. Looking at the overview of the identified trends of the Joint Research Center [104] and the reviews of European distribution system operators [105], it is easy to conclude that innovative scientific research should provide answers to the challenges observed in the literature.

The issue of active distribution network management is observed in recent scientific papers, mostly continuously for the past two years, and factors that can contribute to significant DG integration are identified and structured as multi-objective planning and active management [50,106–114]. Modern grids imply multiple technologies such as advanced metering infrastructure, EMS, DMS, Big Data solutions and data management, pricing mechanisms on a day ahead market and inner day markets, various paradigms such as demand response, fault classification and DG scheduling, and horizontal and/or vertical control possibilities [53,115–117]. Colmenar-Santos et al. [75] detailed the challenges raised from increased DG integration in distribution network identifying advantages and technology characteristics of active management network with DG operational optimization. In the fourth section of their paper authors distinguish optimization techniques based on Reference [54] and give comprehensive overview of advantages and disadvantages of different DG optimization techniques focusing on hybrid methods, thus becoming basis of future research by many scientists. Multi-objective planning, controllable loads, dispatched DG, stochastic DG and demand response become more prominent in Reference [75] as the report progresses. The authors touch upon regulatory and policy barriers preventing significant DG integration in the fifth and sixth sections of their paper, identifying the need for policy change according to the example of Germany, Denmark and Spain [75].

Power System of the future will enable justified use of energy, fairness in energy sharing, security of data and data governance, user own control and autonomy of the household while participating in a fair manner in energy community and virtual energy hubs [118]. The transition of the power industry will not go smoothly and synergy of all actors, legal and technical, regulatory and market, is needed for the transition to succeed [119], but when it does, the benefit will be for everyone included. As mentioned in Section 2, shifting the Smart Grid towards Cloud computing enables smoother transition, and the emergence of the Edge Computing paradigm enables the power industry to flourish [120].

Future of distribution networks are discussed by Bayod-Rújula et al. [50] and authors conclude that active distribution network with total control represents the first stage towards more advanced Smart Grid. According to the authors, second stage of distribution network development represents the microgrid system, a fully controlled entity that may provide ancillary services to the main system; third stage are virtual utilities with operational energy management system yielding optimization opportunities; demand-side management and demand response techniques become available once the previous prerequisites are met hence paving the way towards information and communication technology implementation [50].

In the wake of such information is paper by Shi et al. [109] where authors show operational energy management in microgrid system consisting of dispatchable and non-dispatchable DG, interruptible and deferrable loads and energy storages. Simplified overview of the described system is given by Figure 9, and similar models can be used to validate any other control solution intended for production purposes, as it was the case for the validation purposes of this paper.

Authors comply their proposed management solution with IEC 61850, a communication standard for Smart Grids [121] and present output schedules for grid-connected and islanded modes of operation, respecting bus voltages while minimizing network requirements. Operational energy management in distribution network is identified as the central problem of future distribution networks, microgrids and Smart Grids. Importance of this paper is evident in direct application of the scientific method in real-life microgrid environment. The authors present the results on a daily diagram, divided in 24 intervals, and give operating schedules for PV system, Wind power plant, Diesel generator, battery storage and controllable load.

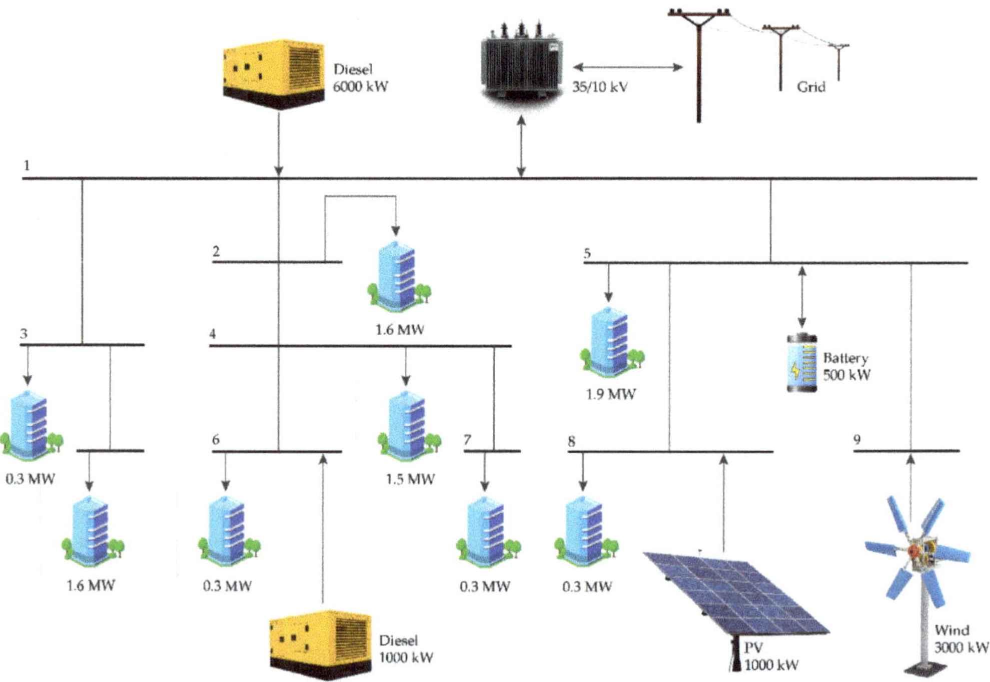

Figure 9. Simplified model of microgrid used for validation in this paper, inspired by Reference [109].

The system is controlled by Micro-Grid Central Controller (MGCC) with Exact, Convex Programming–based algorithm for Nonlinear Systems and Models based on predictor corrector proximal multiplier algorithm and optimal power flow formulation [109].

Operational microgrid energy management based on mixed-integer linear programming solution considering wind power and PV system generation volatile characteristics, aided by storage system and grid connection can be found in the paper by Umeozor and Trifkovic [122]. The authors collected meteorological data to eliminate forecasting errors and schedule DG production accordingly. Different pricing scenarios were considered, and the authors concluded that economic circumstances of observed system greatly affect the output parameters.

The authors of Reference [123] represent the hybrid method emerged by combination of sorting GA without dominant solutions and Rough Set Theory tests. The method is used for solving the problem of optimal energy management in a distribution network that consists of 123 nodes, 7 micro-grid systems and 9 distributed energy sources. The authors suggest a game theory interactive matrix described by 10 mathematical expressions for solving the interaction between different sources observed through the point of common coupling (PCC). Multiple sources are observed during the operative control of one or multiple micro-grid systems inside of distribution network with L number of nodes. Interactive matrix is optimized by the minimization of three objective functions, (26)–(28), using the adapted hierarchically arranged GA, and the proposed method is tested on the IEEE 33 node test network in which the authors integrate three microgrid systems. In expressions (22)–(24), $U_{l,t}$ implies voltage at node l within a network consisting of nodes set L; $P_{l,t}^{PCC}$ stands for PCC power exchange of microgrid at node l; P_t^{LOSS} represents power loss of ADN and θ is power exchange level of the observed system.

$$\min G_1 = \sum_{t=1}^{T} \sqrt{\sum_{l \in L} (U_{l,t} - 1)^2 / L} \tag{26}$$

$$\min G_2 = \sum_{t=1}^{T} \sqrt{\sum_{l \in L} \left(P_{l,t}^{PCC} - \theta\right)^2 / T} \tag{27}$$

$$\min G_3 = \sum_{t=1}^{T} P_t^{LOSS} \tag{28}$$

By comparing the results from the soft computing optimization methods with the results from the classical mathematical approach, the authors conclude that proposed approach yields betters results during the optimization of more complex systems with multiple DG units. The authors emphasize the specific achievement of the paper in the efficient methodology for solving complex optimization problems by using the computer intelligence, while the additional procedures and other hybrid solutions must be explored.

Automatic generation control (AGC) of distribution network with high share of DG and electric vehicles based on framework established for transmission systems is presented in the work by Batisetelli et al. [113]. The authors scaled the known AGC hierarchy to control DG, local distribution and load subsystems and logically divided active distribution network to virtual power isles managed by centralized control. Electric vehicles may contain both generation and load characteristics therefore authors deal with both problems simultaneously through linear programming model optimized by CPLEX optimizer that can be implemented in energy management system of active distribution network. The model described in Reference [113] was tested and confirmed on a realistic four-feeder system from which the "smart user grid (SUG)" paradigm is developed by integrating multi-agent system procedures and entities of different functionalities. Extremely revealing contribution of the authors [113] is evident in "criteria for implementation and coordination of automatic generation control", comprising the communication necessary between observed levels, similar to the one presented in Reference [29]. Discrete 24-segment simulations representing 24-hour schedule is the methodology of most authors, including the one in Reference [113]. Although not the typical computational intelligence approach, paper considered important aspects of ADN management with increased ratio of various types of DG. The absence of advanced optimization techniques is completely justified by an excellent principle demonstration that was clearly the goal of Reference [113], while the practicality and robustness of optimization solution developed on this research basis should consider more complex methods.

Scientific work in the wake of the abovementioned paper can be found in paper by Jun et al. [114] where authors solved the multi-agent optimization problem by means of integer programming dividing the day planning in 24 segments. EMS has a significant role in valid operation, optimization, control and balancing of "hybrid renewable energy generation system", as stated by the authors, and specific challenges of management scheme are described in this paper. Complex unified modeling language diagram, solved in Java Agent Development Framework, of multi-agent system is presented, showing system classes, attributes and relationships between them. The authors introduce new parameters computed by individual agent that indicate the attention of observed agent as load, or as DG. These parameters suggest the economic and power qualities of each agent integrating the stochastic nature of some RES. Detailed analysis of EMS conclude the structural requirements of future communication, behavior and optimizations objectives.

Significant papers indicate the importance of considering meteorological dependencies of some RES-based DGs along with independencies of other types of non-RES DG units. Gu et al. [124] thoroughly defined factors of the future management service consisting of demand side management, operation objectives, energy pricing, weather forecast, energy management, maintenance scheduling, physical limitations of equipment involved, operation scheduling and interaction with main supply grid via PCC. The authors define the perspective of energy management by achieving higher level of efficiency and lower emission levels while maintaining power quality and providing ancillary services.

From their work, one can get a clear idea of what a future energy management system for the distribution network should look like and what should be taken into account, as illustrated by Figure 10. The authors thus state how to respect consumers, plan according to meteorological indicators, consider equipment limitations and the situation on the energy market, all in order to plan the optimal timetable of DG.

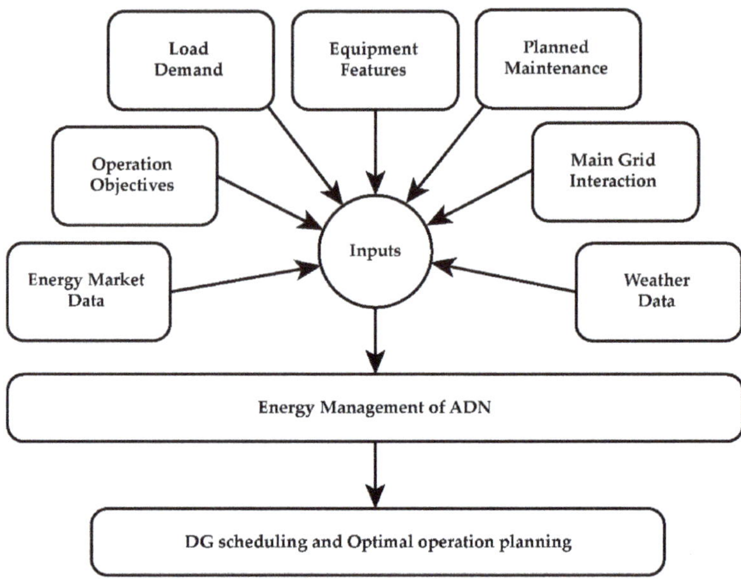

Figure 10. Energy management system of the future distribution grids, inspired by Gu et al.

Zhang et al. [125] concluded that operational management can reduce line losses and improve voltage profiles which will benefit DG developers as well as distribution network operators. The authors define the traditional distribution network planning and operation as "fit and forget" methodology and conclude that this principle will need to change in order to allow higher penetration levels and development of Smart Grid. Mohamed and Mohammed [126] developed effective mathematical algorithm for distribution grid management in a Smart Grid form by giving priority to renewable energy resources while maintaining least possible cost and satisfying load demand. The authors predicted PV generation by historical data and statistical smoothing techniques, while wind generation and load data were modeled by non-linear regression modeling. The main objective achieved by Reference [126] is minimizing the power obtained from parent grid while retaining additional power for sudden demand changes. A case study of Smart Grid management in Japan is presented in Reference [127] from a policy perspective, and an essential requirement for management services in order to achieve maximal benefits is highlighted in this paper.

Significant achievement towards real-time economic dispatch and operational management of active distribution network is evident in paper by Kellerer et al. [128] in which authors propose new method based on statistical inference method—a probabilistic graphical model method. The authors prove algorithm functionality in global optimal search, irrelevant of network size, and with the assumptions that 70% of consumer nodes have randomly integrated RES-based DG units. The main subject of their observations is in solving the economic dispatch problems without having a full knowledge of all component models, for which they assume is difficult in competitive market environment. The proposed algorithm is successfully tested on a network whose topology is based on real Smart Grid data in Southern Germany.

With the objective of reducing CO_2 emissions Lamadrid et al. [129] proposed a novel optimization solution that favors RES in power dispatch and scheduling, if the technology permits planning. From the perspective of system operator and in compliance with network policies, authors present their optimization formulation respecting technical and economic limitations and prove the usefulness of their method by testing it on 279-bus transmission network in Texas, USA. The authors have taken into consideration a stochastic nature of some RES and gave them priority in dispatch schedule. Although modeling is complex, authors managed to perform it correctly by using mathematical correct workaround solutions. By proposed solution authors managed to simulate a future 24-hour period which is adequate for system schedule and operation in market conditions. For wind power modeling, authors used combination of historical and calculated data. This paper is very important due to the complexity of observed problem, and although the authors themselves say that improvements are needed, this paper presents state-of-the-art in operational management and DG scheduling.

A modified self-adaptive PSO algorithm serviceability for multi-objective optimal operation management of a distribution network containing a fuel-cell power plant was proven in paper by Niknam et al. [130], and a computational intelligence principle was described that contained fuzzy-controlled decision-making. The authors developed a practical 24-segment algorithm, providing Pareto solutions. A similar fuzzy controlled approach can be found in paper by Elamine et al. [131] where wind speed is determined by PSO optimized ANN based on small historical set and fuzzy agent for battery storage management.

Previous scientific works have successfully identified key issues and constraints that need to be addressed when creating a widely applicable solution for active distribution network operational management and scheduling. The biggest disadvantage of presented papers is the need for a complex mathematical model of the observed system which is usually not sensitive to structural changes in the same system, hence not applicable to distribution grids. Accordingly, universal applicability of the proposed solutions is not always achieved, and each observed case is at the same time the only case observed and researched by a group of scientists. Issues in Smart Grid application can be compartmentalized into five categories [132]: concept presentation—advantages and limitations of renewable energy distributed generation; technology adoption for hybrid energy system; optimal allocation problems and technical characteristics; forecasting, pricing and policy issues; Smart Grid–integration challenges.

With the development of active distribution networks, which aspire to become Smart Grids, the emphasis of scientists in this area will tend to be on the development of widely applicable and robust solutions for which it will be necessary to consume and respect the knowledge and achievements displayed in all of the so-far listed scientific works.

6. Conclusions

The initial premise of the presented research was to identify scientific breakthroughs and papers that can be delivered as a working solution for future power grids.

The research was conceived in five key steps that begin with the selection of papers and continues with the modeling of the described procedures and testing of the solutions traceability. Finally, the usefulness and applicability were discussed at the research team level, and conclusions are given in this paper.

Papers with precise models and comprehensive descriptions of computational intelligence–based methods proved to be highly usable in the development of real solutions for Smart Grid operational management. Modern systems for monitoring and managing an advanced distribution network according to the assumed European Smart Grid framework will have to integrate scientifically validated and model-tested solutions that consistently provide accurate data.

Within the space–time possibilities and within the available literature, the authors of this paper collected, processed, examined and evaluated as many papers as possible, which with their content promise applicability of computational intelligence in real-world systems. If any work was missed, it was certainly not intentional, and the knowledge and information presented in this article were compiled with the idea of serving anyone involved in optimization and management in the Smart Grid environment. The presented works systematically and precisely present the advantages of the applicability of metaheuristic methods in the optimization of various challenges in the power industry, with an emphasis on the optimization of operating conditions in the distribution network. Although the emphasis of the paper was placed on metaheuristics—because only with it does it become possible to solve very complex systems—the paper also presents simpler analytical methods that can be used to solve the optimization problems of smaller systems. However, scalability of potential real-world solutions can only be ensured with computational intelligence and advanced custom-made metaheuristics.

This paper illustrates the problem of planning the optimal operation of an advanced distribution network with a significant number of distributed generation units and provides the cognitive process of deciding on the framework for the development of contemporary management solutions. Metaheuristic proves to be a necessity in large and complex systems, such as the modern Smart Grid, and modern control solution that integrate metaheuristic methods will provide good-enough solutions for everyday usage. Operators of the future grid will need to settle with such "good-enough" solutions as the grid becomes more complex and variable in nature.

A detailed overview of more than a hundred key papers with important mathematical indicators provides a unique insight into the application of computer intelligence in power engineering. According to the authors of this paper, population methods in a limited search space can yield the best result in terms of repeatability and usability, and with increasing computing power, such optimization procedures based on computational intelligence can take place within the market conditions of real systems.

Finally, methods, procedures and paradigms that can provide reliable operational management in advanced distribution networks were explained, and features of each of them were discussed with the mark-up of leading algorithms, PSO and GA, which can be, to a greater or lesser extent, applied in real-world applications.

Author Contributions: Conceptualization, M.V. and P.M.; methodology, M.V. and G.H.; validation, Z.B. and S.S.; formal analysis, Z.B.; resources, P.M.; writing—original draft preparation, M.V.; writing—review and editing, M.V. and P.M.; visualization, M.V. and G.H.; revision, M.V. and P.M. All authors have read and agreed to the published version of the manuscript.

Funding: This research was co-funded by project "PEGASOS", through the grant of Croatian Ministry of Regional Development and EU Funds "Increasing the development of the new products and services arising from research and development activities—phase II/KK.01.2.1.02" of the program "Competitiveness and Cohesion 2014–2020" from the European Regional Development Fund. The information and views set out in this study are those of the author(s) and do not necessarily reflect the official opinion of the European Union and national implementing bodies. Neither the European Union institutions and bodies nor any person acting on their behalf may be held responsible for the use which may be made of the information contained therein.

Acknowledgments: The authors of this paper express their deepest gratitude to the external domain experts from the Faculty of Electrical Engineering, Computer Science and Information Technology Osijek. Special appreciation authors declare to the experts from Končar—Power Plant and Electric Traction Engineering, Inc., who contributed to validity and analysis. Finally, the authors are grateful to the whole team of Base58 for their technical virtuosity in modeling and testing, administrative skills in project preparation and management, and finally, their motivation and support.

Conflicts of Interest: The authors declare no conflict of interest. The funders had no role in the design of the study; in the collection, analyses or interpretation of data; in the writing of the manuscript; or in the decision to publish the results.

Appendix A

For better understanding and reproduction of the findings of this paper, and for easier catching-up with the mentioned papers, the authors have prepared an abbreviation list. Since some authors choose similar letters and abbreviations, in the description field of Table A1, we give the paper to which the abbreviation applies.

Table A1. Nomenclature and abbreviation list.

Abbreviation	Description	Unit
α_{ij}	Sensitivity factor of real power loss with respect to Real power injection from DG (Acharya et al., 2006)	MW
α_{ij}	The transmission coefficient of the real part of the complex power by Kansal et al. (2013)	n
A_{ij}	Quotient of product of voltage angle difference cosine and line resistance with voltage level (Biswas et al., 2012)	
β_{ij}	Sensitivity factor of reactive power loss with respect to reactive power injection from DG (Acharya et al., 2006)	MVAr
β_{ij}	The transmission coefficient of the reactive part of the complex power by Kansal et al. (2013)	n
B_{ij}	Quotient of product of voltage angle difference sinus and line resistance with voltage level (Biswas et al., 2012)	
β_1	Penalty coefficient ($\beta_1 = 0.32$) (Moradi and Abedini, 2012)	$\beta_1 = 0.32$
β_2	Penalty coefficient ($\beta_2 = 0.3$) (Moradi and Abedini, 2012)	$\beta_2 = 0.3$
C_i^a	Active power prices with DG (Singh and Goswami, 2010)	US$/MWh
C_i^r	Reactive power prices with DG (Singh and Goswami, 2010)	US$/MWh
C_{DG}	Total cost associated with the DGs (Biswas et al., 2012)	US$ mil.
C_{fc}	Cost of power generation from fuel cell (Niknam et al., 2011)	US$
C_{wind}	Cost of power generation from wind power (Niknam et al., 2011)	US$
C_{pv}	Cost of power generation from photovoltaic system (Niknam et al., 2011)	US$
$Cost_{sub}$	Cost of substation (Niknam et al., 2011)	US$
δ_{ij}	Voltage angle difference between bus i and bus j (Alrashidi and Alhajri, 2011)	rad
$E_{t_{fc}}$	Emission of atmospheric pollutants from fuel-cell power generation (Niknam et al., 2011)	kg/h
$E_{t_{wind}}$	Emission of atmospheric pollutants from wind power generation (Niknam et al., 2011)	kg/h
$E_{t_{pv}}$	Emission of atmospheric pollutants from photovoltaic power generation (Niknam et al., 2011)	kg/h
I_i	Actual current of the i-th branch (Niknam et al., 2011)	A
ILP	Real power losses after DG integration by El-Zonkoly (2014.)	%
ILQ	Reactive power losses after DG integration by El-Zonkoly (2014.)	%
IC	MVA capacity index regarding the power flows through conductors in paper by El-Zonkoly (2014.)	%
IVD	Voltage profile index, observed as voltage deviation from the nominal value by El-Zonkoly (2014.)	%
ISC	Short circuit level index, observed as DG impact on short circuit current increase in paper by El-Zonkoly (2014.)	%
k_1	Penalty coefficient (Moradi and Abedini, 2012)	$k_1 = 0.6$

Table A1. *Cont.*

Abbreviation	Description	Unit
k_2	Penalty coefficient (Moradi and Abedini, 2012)	$k_2 = 0.35$
K_C	Is the cost of DG per KW (Biswas et al., 2012)	US$
λ	Electricity price at power supply point (Singh and Goswami, 2010)	US$/MWh
L	Set of nodes in ADN by Lv and Ai (2016)	n
LLR	Total line-loss reduction (Abou El-Ela et al., 2010)	%
$LL_{wo/DG}$	Line-loss without the DG (Abou El-Ela et al., 2010)	MW
$LL_{w/DG}$	Line-loss with the DG (Abou El-Ela et al., 2010)	MW
L_{DIST_i}	Load distributed for the i-th fault (Biswas et al., 2012)	MW
$MBDG$	Maximal composite benefits of DG (Abou El-Ela et al., 2010)	
MOF	PSO-based multi-objective function by El-Zonkoly (2014.)	
N	Number of buses in distribution system by Kansal et al.	N
NB	Number of radial distribution system buses (Alrashidi and Alhajri, 2011)	n
N_{br}	Number of the branches (Niknam et al., 2011)	n
N_{bus}	Total number of the buses (Niknam et al., 2011)	n
N_d	Number of years (Niknam et al., 2011)	n
N_F	Total number of faults within a specified time duration (Biswas et al., 2012)	
N_{fc}	Number of fuel-cell units (Niknam et al., 2011)	n
N_{pv}	Number of photovoltaic units (Niknam et al., 2011)	n
N_{wind}	Number of wind units (Niknam et al., 2011)	n
θ	Power exchange level by Lv and Ai (2016)	kW/t
P_{ω}^{t}	Inertial probability (Gomez-Gonzalez et al., 2012)	$Q \in [0, 1]$
$p_{\omega,max}$	Maximal inertia per population (Gomez-Gonzalez et al., 2012)	$Q \in [0, 1]$
$p_{\omega,min}$	Minimal inertia per population (Gomez-Gonzalez et al., 2012)	$Q \in [0, 1]$
P_{Di}	Active power demand at any bus i (Singh and Goswami, 2010)	MW
P_{DGi}	Real power injection at node i (Acharya et al., 2006)	MW
P_{DGi}	Active power generated by DG (Singh and Goswami, 2010)	MW
P_{DG_i}	Size of the i-th DG (Biswas et al., 2012)	MW
P_{Di}	Load demand at node i (Acharya et al., 2006)	MW
P_i^{f2}	Real power flows at the receiving end of the forward-update procedure by Injeti el al.	MW
P_i, P_j	The active power injections at bus i and bus j by Kansal et al.	MW
P_{fc}	Power generated from fuel cell (Niknam et al., 2011)	kWh
P_G	Active power of distributed generation (Aman et al., 2012)	MW
$\mathrm{P}_{1,t}^{\mathrm{PCC}}$	PCC power exchange of MG n by Lv and Ai (2016)	kW
P_{Loss}	Line losses (Injeti and Prema Kumar, 2013)	MW
$P_{T,Loss}$	Total feeder losses (Injeti and Prema Kumar, 2013)	MW
$\mathrm{P}_{t}^{\mathrm{LOSS}}$	power loss of ADN in paper by Lv and Ai (2016)	kW
P_{eleect}^{DG}	The price of electricity (Singh and Goswami, 2010)	US$/MWh
PFR	Power flow reduction in critical lines (Abou El-Ela et al., 2010)	%
$PF_{k,wo/DG}$	Power flow in line k without DG (Abou El-Ela et al., 2010)	MW

Table A1. *Cont.*

Abbreviation	Description	Unit
$PF_{k,w/DG}$	Power flow in line k with DG (Abou El-Ela et al., 2010)	MW
P_{pv}	Power generated from photovoltaic system (Niknam et al., 2011)	kWh
P_{wind}	Power generated from wind power (Niknam et al., 2011)	kWh
$Q_i^{\prime 2}$	Reactive power flows at the receiving end of the forward-update procedure by Injeti el al.	MVAr
Q_{Di}	Reactive power demand at any bus i (Singh and Goswami, 2010)	MVAr
Q_{DGi}	Reactive power generated by DG (Singh and Goswami, 2010)	MVAr
Q_i, Q_j	The reactive power injections at bus i and bus j by Kansal et al.	MVAr
R_i	Resistance of the i-th branch (Niknam et al., 2011)	Ω/km
r_{ij}	Real part of line impedance (Aman et al., 2012)	Ω/km
$R_{1i}(j)$	Equivalent resistance between bus 1 and bus i when DG is located at bus j (Wang and Nehrir, 2004)	p.u.
RPL	Real power loss (Biswas et al., 2012)	MW
σ_1, σ_2, σ_3, σ_4, σ_5	Weights for corresponding importance of each DG impact index by El-Zonkoly (2014.)	$\sigma_p \in [0, 1]$
S_{DIST}	Total load distributed (Biswas et al., 2012)	
$\|S_{ni}\|$	Apparent power at bus n_i(Moradi and Abedini, 2012)	MVA
$\|S_{ni}^{max}\|$	Maximum apparent power at bus n_i(Moradi and Abedini, 2012)	MVA
$SR_{w/DG}$	Spinning reserve with DG [23] (Abou El-Ela et al., 2010)	p.u.
$SR_{wo/DG}$	Spinning reserve without DG (Abou El-Ela et al., 2010)	p.u.
SRI	Spinning reserve increasing (Abou El-Ela et al., 2010)	%
T	The sum of scheduling period t in paper by Lv and Ai (2016)	n
Δt	Time step (one year) (Niknam et al., 2011)	n
U_n	set of types of DGs in Microgrid n; u = 1, 2, 3, 4 and 5 denote fuel cell, microturbine, battery storage, Photovoltaic System and Wind turbine, respectively, by Lv and Ai (2016)	$n \in [0, 5]$
V_i	The bus i voltage (Alrashidi and Alhajri, 2011)	p.u.
V_i	Real voltage of the i-th bus (Niknam et al., 2011)	p.u.
$\|V_i\|^2$	Voltage magnitude at the receiving end of the forward-update procedure Injeti el al.	p.u.
V_j	The bus j voltage (Alrashidi and Alhajri, 2011)	p.u.
V_{ni}	Voltage of bus n_i(Moradi and Abedini, 2012)	p.u.
V_{ni}^{max}	Maximum voltage at bus n_i(Moradi and Abedini, 2012)	p.u.
V_{ni}^{min}	Minimum voltage at bus n_i [(Moradi and Abedini, 2012)	p.u.
V_{Rating}, V_{rate}	Nominal voltage of the i-th bus (Niknam et al., 2011)	p.u.
VPI	Voltage profile improvement (Abou El-Ela et al., 2010)	%
$VP_{w/DG}$	Voltage profile index of the system with DG (Abou El-Ela et al., 2010)	p.u.
$VP_{wo/DG}$	Voltage profile index without DG (Abou El-Ela et al., 2010)	p.u.
w_1, w_2	Weight factors, $w_1 + w_2 = 1$(Yang and Chen, 2011)	
w_1, w_2, w_3, w_4	Benefit weighting factors (Abou El-Ela et al., 2010)	

References

1. Colak, I.; Fulli, G.; Sagiroglu, S.; Yesilbudak, M.; Covrig, C.F. Smart grid projects in Europe: Current status, maturity and future scenarios. *Appl. Energy* **2015**, *152*, 58–70. [CrossRef]
2. Palensky, P.; van der Meer, A.A.; Lopez, C.D.; Joseph, A.; Pan, K. Cosimulation of Intelligent Power Systems: Fundamentals, Software Architecture, Numerics, and Coupling. *IEEE Ind. Electron. Mag.* **2017**, *11*, 34–50. [CrossRef]

3. Pérez-Ortiz, M.; Jiménez-Fernández, S.; Gutiérrez, P.A.; Alexandre, E.; Hervás-Martínez, C.; Salcedo-Sanz, S. A Review of Classification Problems and Algorithms in Renewable Energy Applications. *Energies* **2016**, *9*, 607. [CrossRef]
4. Uslar, M.; Schmedes, T.; Lucks, A.; Luhmann, T.; Winkels, L.; Appelrath, H.J. Interaction of EMS related systems by using the CIM standard. In Proceedings of the ITEE 2005 2nd Internacional ICSC Symposium Information Technologies Environmental Engineering, Magdenburg, Germany, 25–27 September 2005; pp. 596–610.
5. Molzahn, D.K.; Dorfler, F.; Sandberg, H.; Low, S.H.; Chakrabarti, S.; Baldick, R.; Lavaei, J. A Survey of Distributed Optimization and Control Algorithms for Electric Power Systems. *IEEE Trans. Smart Grid* **2017**, *8*, 2941–2962. [CrossRef]
6. Liao, H. Review on Distribution Network Optimization under Uncertainty. *Energies* **2019**, *12*, 3369. [CrossRef]
7. Theo, W.L.; Lim, J.S.; Ho, W.S.; Hashim, H.; Lee, C.T. Review of distributed generation (DG) system planning and optimisation techniques: Comparison of numerical and mathematical modelling methods. *Renew. Sustain. Energy Rev.* **2017**, *67*, 531–573. [CrossRef]
8. Mahmoud Pesaran, H.A.; Huy, P.D.; Ramachandaramurthy, V.K. A review of the optimal allocation of distributed generation: Objectives, constraints, methods, and algorithms. *Renew. Sustain. Energy Rev.* **2016**, *75*, 293–312. [CrossRef]
9. Ruiz-Romero, S.; Colmenar-Santos, A.; Mur-Pérez, F.; López-Rey, Á. Integration of distributed generation in the power distribution network: The need for smart grid control systems, communication and equipment for a smart city—Use cases. *Renew. Sustain. Energy Rev.* **2014**, *38*, 223–234. [CrossRef]
10. Madsen, D.Ø. The Evolutionary Trajectory of the Agile Concept Viewed from a Management Fashion Perspective. *Soc. Sci.* **2020**, *9*, 69. [CrossRef]
11. Malik, M.; Sarwar, S.; Orr, S. Agile practices and performance: Examining the role of psychological empowerment. *Int. J. Proj. Manag.* **2021**, *39*, 10–20. [CrossRef]
12. Viral, R.; Khatod, D. Optimal planning of distributed generation systems in distribution system: A review. *Renew. Sustain. Energy Rev.* **2012**, *16*, 5146–5165. [CrossRef]
13. Sambaiah, K.S. A review on optimal allocation and sizing techniques for DG in distribution systems. *Int. J. Renew. Energy Res.* **2018**, *8*, 1236–1256.
14. Mazidi, M.; Zakariazadeh, A.; Jadid, S.; Siano, P. Integrated scheduling of renewable generation and demand response programs in a microgrid. *Energy Convers. Manag.* **2014**, *86*, 1118–1127. [CrossRef]
15. Lopes, J.P.; Hatziargyriou, N.; Mutale, J.; Djapic, P.; Jenkins, N. Integrating distributed generation into electric power systems: A review of drivers, challenges and opportunities. *Electr. Power Syst. Res.* **2007**, *77*, 1189–1203. [CrossRef]
16. Kim, H.-S.; Hong, J.; Choi, I.-S. Implementation of Distributed Autonomous Control Based Battery Energy Storage System for Frequency Regulation. *Energies* **2021**, *14*, 2672. [CrossRef]
17. Urbanetz, J.; Braun, P.; Rüther, R. Power quality analysis of grid-connected solar photovoltaic generators in Brazil. *Energy Convers. Manag.* **2012**, *64*, 8–14. [CrossRef]
18. Yang, N.-C.; Chen, T.-H. Evaluation of maximum allowable capacity of distributed generations connected to a distribution grid by dual genetic algorithm. *Energy Build.* **2011**, *43*, 3044–3052. [CrossRef]
19. Vukobratović, M.; Nikolovski, S.; Marić, P. Improving the Conditions in a Radial Distribution Feeder by Implementing Distributed Generation. *Int. J. Electr. Comput. Eng. Syst.* **2021**, *6*, 5. [CrossRef]
20. Ipinnimo, O.; Chowdhury, S.; Mitra, J. A review of voltage dip mitigation techniques with distributed generation in electricity networks. *Electr. Power Syst. Res.* **2013**, *103*, 28–36. [CrossRef]
21. Yadav, A.; Srivastava, L. Optimal placement of distributed generation: An overview and key issues. In Proceedings of the 2014 International Conference on Power Signals Control and Computations (EPSCICON), Atlanta, GA, USA, 3–5 December 2014; IEEE: Piscataway, NJ, USA, 2014; pp. 8–10. [CrossRef]
22. Djafar, W.; Amer, Y.; Lee, S.-H. Models and Optimisation Techniques on Long Distribution Network: A Review. *Procedia Manuf.* **2015**, *2*, 519–526. [CrossRef]
23. Kotamarty, S.; Khushalani, S.; Schulz, N. Impact of distributed generation on distribution contingency analysis. *Electr. Power Syst. Res.* **2008**, *78*, 1537–1545. [CrossRef]
24. Bignucolo, F.; Cerretti, A.; Coppo, M.; Savio, A.; Turri, R. Impact of Distributed Generation Grid Code Requirements on Islanding Detection in LV Networks. *Energies* **2017**, *10*, 156. [CrossRef]
25. Acharya, N.; Mahat, P.; Mithulananthan, N. An analytical approach for DG allocation in primary distribution network. *Int. J. Electr. Power Energy Syst.* **2006**, *28*, 669–678. [CrossRef]
26. Gözel, T.; Hocaoglu, M.H. An analytical method for the sizing and siting of distributed generators in radial systems. *Electr. Power Syst. Res.* **2009**, *79*, 912–918. [CrossRef]
27. Injeti, S.K.; Kumar, N.P. A novel approach to identify optimal access point and capacity of multiple DGs in a small, medium and large scale radial distribution systems. *Int. J. Electr. Power Energy Syst.* **2013**, *45*, 142–151. [CrossRef]
28. Bollen, M.; Hassan, F. *Integration of Distributed Generation in the Power System*; Wiley: Hoboken, NJ, USA, 2011; ISBN 978-1-118-02901-5.
29. Bruinenberg, J.; Colton, L.; Darmois, E.; Dorn, J.; Doyle, J.; Elloumi, O.; Englert, H.; Forbes, R.; Heiles, J.; Hermans, P.; et al. *CEN -CENELEC—ETSI: Smart Grid Coordination Group—Smart Grid Reference Architecture Report 2.0.*; CEN-CENELEC-ETSI: Brussels, Belgium, 2012.

30. Shewale, A.; Mokhade, A.; Funde, N.; Bokde, N.D. An Overview of Demand Response in Smart Grid and Optimization Techniques for Efficient Residential Appliance Scheduling Problem. *Energies* **2020**, *13*, 4266. [CrossRef]
31. Panda, D.K.; Das, S. Smart grid architecture model for control, optimization and data analytics of future power networks with more renewable energy. *J. Clean. Prod.* **2021**, *301*, 126877. [CrossRef]
32. Markovic, D.S.; Zivkovic, D.; Branovic, I.; Popovic, R.; Cvetkovic, D. Smart power grid and cloud computing. *Renew. Sustain. Energy Rev.* **2013**, *24*, 566–577. [CrossRef]
33. Machinda, G.T.; Chowdhury, S.; Mbav, W.N. Power management of inverter interfaced solar PV microgrid: A review of the current technological trend. In Proceedings of the 47th International Universities Power Engineering Conference (UPEC), London, UK, 4–7 September 2012; pp. 1–6.
34. Choudar, A.; Boukhetala, D.; Barkat, S.; Brucker, J.-M. A local energy management of a hybrid PV-storage based distributed generation for microgrids. *Energy Convers. Manag.* **2015**, *90*, 21–33. [CrossRef]
35. Häger, U.; Rehtanz, C.; Editors, N.V. *Power Systems Monitoring, Control and Protection of Interconnected Power Systems*; Springer: Berlin/Heidelberg, Germany, 2014; ISBN 978-3-642-53847-6.
36. Adefarati, T.; Bansal, R.C. Integration of renewable distributed generators into the distribution system: A review. *IET Renew. Power Gener.* **2016**, *10*, 873–884. [CrossRef]
37. Wang, C.; Nehrir, M.H.H. Analytical Approaches for Optimal Placement of Distributed Generation Sources in Power Systems. *IEEE Trans. Power Syst.* **2004**, *19*, 2068–2076. [CrossRef]
38. Aman, M.M.; Jasmon, G.B.; Mokhlis, H.; Bakar, A.H.A. Optimal placement and sizing of a DG based on a new power stability index and line losses. *Int. J. Electr. Power Energy Syst.* **2012**, *43*, 1296–1304. [CrossRef]
39. Singh, R.K.; Goswami, S.K. Optimum allocation of distributed generations based on nodal pricing for profit, loss reduction, and voltage improvement including voltage rise issue. *Int. J. Electr. Power Energy Syst.* **2010**, *32*, 637–644. [CrossRef]
40. El-Ela, A.A.; Allam, S.M.; Shatla, M.M. Maximal optimal benefits of distributed generation using genetic algorithms. *Electr. Power Syst. Res.* **2010**, *80*, 869–877. [CrossRef]
41. Biswas, S.; Goswami, S.K.; Chatterjee, A. Optimum distributed generation placement with voltage sag effect minimization. *Energy Convers. Manag.* **2012**, *53*, 163–174. [CrossRef]
42. AlRashidi, M.R.; Alhajri, M.F. Optimal planning of multiple distributed generation sources in distribution networks: A new approach. *Energy Convers. Manag.* **2011**, *52*, 3301–3308. [CrossRef]
43. González, M.G.; López, A.; Jurado, F. Optimization of distributed generation systems using a new discrete PSO and OPF. *Electr. Power Syst. Res.* **2012**, *84*, 174–180. [CrossRef]
44. Moradi, M.; Abedini, M. A combination of genetic algorithm and particle swarm optimization for optimal DG location and sizing in distribution systems. *Int. J. Electr. Power Energy Syst.* **2012**, *34*, 66–74. [CrossRef]
45. Soares, J.; Silva, M.; Sousa, T.; Vale, Z.; Morais, H. Distributed energy resource short-term scheduling using Signaled Particle Swarm Optimization. *Energy* **2012**, *42*, 466–476. [CrossRef]
46. Saif, A.; Pandi, V.R.; Zeineldin, H.; Kennedy, S. Optimal allocation of distributed energy resources through simulation-based optimization. *Electr. Power Syst. Res.* **2013**, *104*, 1–8. [CrossRef]
47. Kansal, S.; Kumar, V.; Tyagi, B. Optimal placement of different type of DG sources in distribution networks. *Int. J. Electr. Power Energy Syst.* **2013**, *53*, 752–760. [CrossRef]
48. Hung, D.Q.; Mithulananthan, N.; Lee, K.Y. Optimal placement of dispatchable and nondispatchable renewable DG units in distribution networks for minimizing energy loss. *Int. J. Electr. Power Energy Syst.* **2014**, *55*, 179–186. [CrossRef]
49. Aman, M.; Jasmon, G.B.; Bakar, A.H.A.; Mokhlis, H. A new approach for optimum simultaneous multi-DG distributed generation Units placement and sizing based on maximization of system loadability using HPSO (hybrid particle swarm optimization) algorithm. *Energy* **2014**, *66*, 202–215. [CrossRef]
50. Bayod-Rújula, A.A. Future development of the electricity systems with distributed generation. *Energy* **2009**, *34*, 377–383. [CrossRef]
51. Borges, C.L.T.; Martins, V.F. Multistage expansion planning for active distribution networks under demand and Distributed Generation uncertainties. *Int. J. Electr. Power Energy Syst.* **2012**, *36*, 107–116. [CrossRef]
52. El-Khattam, W.; Salama, M.M.A. Distributed generation technologies, definitions and benefits. *Electr. Power Syst. Res.* **2004**, *71*, 119–128. [CrossRef]
53. Alotaibi, I.; Abido, M.A.; Khalid, M.; Savkin, A.V. A Comprehensive Review of Recent Advances in Smart Grids: A Sustainable Future with Renewable Energy Resources. *Energies* **2020**, *13*, 6269. [CrossRef]
54. Tan, W.-S.; Hassan, M.Y.; Majid, S.; Rahman, H.A. Optimal distributed renewable generation planning: A review of different approaches. *Renew. Sustain. Energy Rev.* **2013**, *18*, 626–645. [CrossRef]
55. Ongsakul, W.; Vo Ngoc, D. *Artificial Intelligence in Power System Optimization*; CRC Press: Boca Raton, FL, USA, 2013; ISBN 978-1-4665-7342-0.
56. Ashlock, D. Evolutionary Computation for Modeling and Optimization. In *Evolutionary Computation for Modeling and Optimization*; Springer: Berlin/Heidelberg, Germany, 2006.
57. Kordon, A. *Applying Computational Intelligence*; Springer: Berlin/Heidelberg, Germany, 2010; ISBN 978-3-540-69910-1.
58. Sumathi, S.; Surekha, P. *Computational Intelligence Paradigms: Theory and Applications Using MATLAB*; CRC Press: Boca Raton, FL, USA, 2010; ISBN 978-1-4398-0902-0.

59. Sörensen, K. Metaheuristics-the metaphor exposed. *Int. Trans. Oper. Res.* **2013**, *22*, 3–18. [CrossRef]
60. Tang, K.S. *Multiobjective Optimization Methodology: A Jumping Gene Approach*; Wilamowski, B.M., Irwin, D.J., Eds.; CRC Press: Boca Raton, FL, USA, 2018; ISBN 978-1-4398-9919-9.
61. Gopalakrishnan, K.; Khaitan, S.K.; Kalogirou, S. *Soft Computing in Green and Renewable Energy Systems*; Gopalakrishnan, K., Khaitan, S.K., Kalogirou, S., Eds.; Studies in Fuzziness and Soft Computing; Springer: Berlin/Heidelberg, Germany, 2011; Volume 269, ISBN 978-3-642-22175-0.
62. Luna-Rubio, R.; Trejo-Perea, M.; Vargas-Vázquez, D.; Ríos-Moreno, G. Optimal sizing of renewable hybrids energy systems: A review of methodologies. *Sol. Energy* **2012**, *86*, 1077–1088. [CrossRef]
63. Rutkowski, L. *Computational Intelligence Methods and Techniques*; Springer: Warszawa, Poland, 2005; ISBN 978-3-540-76287-4.
64. Borenstein, Y.; Moraglio, A. *Theory and Principled Methods for the Design of Metaheuristics*; Borenstein, Y., Moraglio, A., Eds.; Natural Computing Series; Springer: Berlin/Heidelberg, Germany, 2014; ISBN 978-3-642-33205-0.
65. Yang, X.-S.; Chien, S.F.; Ting, T.O. Computational Intelligence and Metaheuristic Algorithms with Applications. *Sci. World J.* **2014**, *2014*, 425853. [CrossRef] [PubMed]
66. Kennedy, J. Swarm Intelligence. In *Handbook of Nature-Inspired and Innovative Computing*; Kluwer Academic Publishers: Boston, UK, 2001; pp. 187–219. ISBN 978-0-387-40532-2.
67. Kruse, R.; Borgelt, C.; Braune, C.; Mostaghim, S.; Steinbrecher, M. *Computational Intelligence*; Texts in Computer Science; Springer London: London, UK, 2016; ISBN 978-1-4471-7294-9.
68. Hopgood, A.A. The State of Artificial Intelligence. In *Advances in Organometallic Chemistry*; Elsevier BV: Amsterdam, The Netherlands, 2005; Volume 65, pp. 1–75. ISBN 0120121654.
69. Ibrahim, D. An Overview of Soft Computing. *Procedia Comput. Sci.* **2016**, *102*, 34–38. [CrossRef]
70. Kumari, M.S.; Priyanka, G.; Sydulu, M. Comparison of Genetic Algorithms and Particle Swarm Optimization for Optimal Power Flow Including FACTS devices. In Proceedings of the 2007 IEEE Lausanne Power Technology, Lausanne, Switzerland, 1–5 July 2007; pp. 1105–1110.
71. Liu, S.; Hou, Z.; Wang, M. A hybrid algorithm for optimal power flow using the chaos optimization and the linear interior point algorithm. In Proceedings of the International Conference Power System Technology, Kunming, China, 13–17 October 2002; Volume 4, pp. 793–797.
72. AlRashidi, M.; El-Hawary, M. Applications of computational intelligence techniques for solving the revived optimal power flow problem. *Electr. Power Syst. Res.* **2009**, *79*, 694–702. [CrossRef]
73. Santofimia-Romero, M.-J.; Toro-García, X.; López-López, J.-C. *Artificial Intelligence Techniques for Smart Grid Applications*; CEPIS: Brussels, Belgium, 2011; Volume XII.
74. Abraham, A.; Das, S. *Computational Intelligence in Power Engineering*; Panigrahi, B.K., Abraham, A., Das, S., Eds.; Studies in Computational Intelligence; Springer: Berlin/Heidelberg, Germany, 2010; Volume 302, ISBN 978-3-642-14012-9.
75. Colmenar-Santos, A.; Reino-Rio, C.; Borge-Diez, D.; Collado-Fernández, E. Distributed generation: A review of factors that can contribute most to achieve a scenario of DG units embedded in the new distribution networks. *Renew. Sustain. Energy Rev.* **2016**, *59*, 1130–1148. [CrossRef]
76. Sotkiewicz, P.M.; Vignolo, J.M. Nodal Pricing for Distribution Networks: Efficient Pricing for Efficiency Enhancing DG. *IEEE Trans. Power Syst.* **2006**, *21*, 1013–1014. [CrossRef]
77. Mitchell, M. *An Introduction to Genetic Algorithms*; The MIT Press: Cambridge, MA, USA; London, UK, 1996; ISBN 0262631857.
78. Poli, R.; Langdon, W.; McPhee, N. *A Field Guide to Genetic Programming*; Lulu Press, Inc.: Morrisville, NC, USA, 2008; ISBN 9781409200734.
79. Haupt, R.L.; Haupt, S.E. *Practical Genetic Algorithms*; Wiley: Hoboken, NJ, USA, 2004; ISBN 0-471-45565-2.
80. Injeti, S.K. A Pareto optimal approach for allocation of distributed generators in radial distribution systems using improved differential search algorithm. *J. Electr. Syst. Inf. Technol.* **2018**, *5*, 908–927. [CrossRef]
81. Shammah, A.; El-Ela, A.A.; Azmy, A.M. Optimal location of remote terminal units in distribution systems using genetic algorithm. *Electr. Power Syst. Res.* **2012**, *89*, 165–170. [CrossRef]
82. Wu, Q.H.; Cao, Y.J.; Wen, J.Y. Optimal reactive power dispatch using an adaptive genetic algorithm. *Int. J. Electr. Power Energy Syst.* **1998**, *20*, 563–569. [CrossRef]
83. Subbaraj, P.; Rajnarayanan, P.N. Optimal reactive power dispatch using self-adaptive real coded genetic algorithm. *Electr. Power Syst. Res.* **2009**, *79*, 374–381. [CrossRef]
84. López-Lezama, J.M.; Contreras, J.; Padilha-Feltrin, A. Location and contract pricing of distributed generation using a genetic algorithm. *Int. J. Electr. Power Energy Syst.* **2012**, *36*, 117–126. [CrossRef]
85. El-Zonkoly, A.M. Optimal placement of multi-distributed generation units including different load models using particle swarm optimization. *Swarm Evol. Comput.* **2011**, *1*, 50–59. [CrossRef]
86. El-Zonkoly, A. Optimal scheduling of observable controlled islands in presence of energy hubs. *Electr. Power Syst. Res.* **2017**, *142*, 141–152. [CrossRef]
87. El-Zonkoly, A. Intelligent energy management of optimally located renewable energy systems incorporating PHEV. *Energy Convers. Manag.* **2014**, *84*, 427–435. [CrossRef]
88. El-Zonkoly, A. Optimal placement and schedule of multiple grid connected hybrid energy systems. *Int. J. Electr. Power Energy Syst.* **2014**, *61*, 239–247. [CrossRef]

89. Aman, M.M.; Jasmon, G.B.; Bakar, A.H.A.; Mokhlis, H. A new approach for optimum DG placement and sizing based on voltage stability maximization and minimization of power losses. *Energy Convers. Manag.* **2013**, *70*, 202–210. [CrossRef]
90. Niknam, T.; Taheri, S.I.; Aghaei, J.; Tabatabaei, S.; Nayeripour, M. A modified honey bee mating optimization algorithm for multiobjective placement of renewable energy resources. *Appl. Energy* **2011**, *88*, 4817–4830. [CrossRef]
91. Niknam, T. An efficient multi-objective HBMO algorithm for distribution feeder reconfiguration. *Expert Syst. Appl.* **2011**, *38*, 2878–2887. [CrossRef]
92. Niknam, T. An efficient hybrid evolutionary algorithm based on PSO and HBMO algorithms for multi-objective Distribution Feeder Reconfiguration. *Energy Convers. Manag.* **2009**, *50*, 2074–2082. [CrossRef]
93. Yin, P.-Y.; Yu, S.-S.; Wang, P.-P.; Wang, Y.-T. A hybrid particle swarm optimization algorithm for optimal task assignment in distributed systems. *Comput. Stand. Interfaces* **2006**, *28*, 441–450. [CrossRef]
94. Zhang, W.; Liu, Y. Multi-objective reactive power and voltage control based on fuzzy optimization strategy and fuzzy adaptive particle swarm. *Int. J. Electr. Power Energy Syst.* **2008**, *30*, 525–532. [CrossRef]
95. Ashari, M.; Soeprijanto, A. Optimal Distributed Generation (DG) Allocation for Losses Reduction Using Improved Particle Swarm Optimization (IPSO) Method. *J. Basic. Appl. Sci. Res.* **2012**, *2*, 7016–7023.
96. Bhumkittipich, K.; Phuangpornpitak, W. Optimal Placement and Sizing of Distributed Generation for Power Loss Reduction Using Particle Swarm Optimization. *Energy Procedia* **2013**, *34*, 307–317. [CrossRef]
97. Zhang, Y.; Wang, S.; Ji, G. A Comprehensive Survey on Particle Swarm Optimization Algorithm and Its Applications. *Math. Probl. Eng.* **2015**, *2015*, 931256. [CrossRef]
98. Moghaddas-Tafreshi, S.M.; Zamani, H.A.; Hakimi, S.M. Optimal sizing of distributed resources in micro grid with loss of power supply probability technology by using breeding particle swarm optimization. *J. Renew. Sustain. Energy* **2011**, *3*, 43105. [CrossRef]
99. Ganguly, S.; Sahoo, N.; Das, D. Multi-objective particle swarm optimization based on fuzzy-Pareto-dominance for possibilistic planning of electrical distribution systems incorporating distributed generation. *Fuzzy Sets Syst.* **2013**, *213*, 47–73. [CrossRef]
100. Vukobratović, M.; Marić, P.; Hederić, Ž. Voltage and power losses control using distributed generation and computational intelligence. *Teh. Vjesn. Tech. Gaz.* **2016**, *23*, 23. [CrossRef]
101. De Tré, G.; Zadrożny, S. *Springer Handbook of Computational Intelligence*; Springer: Berlin/Heidelberg, Germany, 2015; ISBN 9783662435052.
102. EU Commission European. Task Force for the implementation of Smart Grids into the European Internal Market. In *Mission Work Programme*; EU Commission European: Geneva, Switzerland, 2012; pp. 1–15.
103. Joint Research Centre—Smart Electricity Systems and Interoperability European Smart Grid Projects: Relationship Matrix. Available online: http://ses.jrc.ec.europa.eu/european-smart-grid-projects-relationship-matrix (accessed on 12 December 2017).
104. Gangale, F.; Vasiljevska, J.; Covrig, C.F.; Mengolini, A.; Fulli, G. *Smart Grid Projects Outlook 2017: Facts, Figures and Trends in Europe*; Publications Office of the European Union: Luxembourg, 2017.
105. Prettico, G.; Gangale, F.; Mengolini, A.; Lucas, A.; Fulli, G. *DISTRIBUTION SYSTEM OPERATORS OBSERVATORY: From European Electricity Distribution Systems to Representative Distribution Networks*; Publications Office of the European Union: Luxembourg, 2016.
106. Jalali, M.M.; Kazemi, A. Demand side management in a smart grid with multiple electricity suppliers. *Energy* **2015**, *81*, 766–776. [CrossRef]
107. Fahrioglu, M.; Alvarado, F.L.; Lasseter, R.H.; Yong, T. Supplementing demand management programs with distributed generation options. *Electr. Power Syst. Res.* **2012**, *84*, 195–200. [CrossRef]
108. Kinhekar, N.; Padhy, N.P.; Gupta, H.O. Multiobjective demand side management solutions for utilities with peak demand deficit. *Int. J. Electr. Power Energy Syst.* **2014**, *55*, 612–619. [CrossRef]
109. Shi, W.; Xie, X.; Chu, C.-C.; Gadh, R. Distributed Optimal Energy Management in Microgrids. *IEEE Trans. Smart Grid* **2015**, *6*, 1137–1146. [CrossRef]
110. Niknam, T.; Meymand, H.Z.; Nayeripour, M. A practical algorithm for optimal operation management of distribution network including fuel cell power plants. *Renew. Energy* **2010**, *35*, 1696–1714. [CrossRef]
111. Singh, A.; Parida, S. Congestion management with distributed generation and its impact on electricity market. *Int. J. Electr. Power Energy Syst.* **2013**, *48*, 39–47. [CrossRef]
112. Macedo, M.N.Q.; Galo, J.J.M.; de Almeida, L.A.L.; de Lima, A.C. Demand side management using artificial neural networks in a smart grid environment. *Renew. Sustain. Energy Rev.* **2015**, *41*, 128–133. [CrossRef]
113. Battistelli, C.; Conejo, A.J. Optimal management of the automatic generation control service in smart user grids including electric vehicles and distributed resources. *Electr. Power Syst. Res.* **2014**, *111*, 22–31. [CrossRef]
114. Zeng, J.; Liu, J.; Wu, J.; Ngan, H.W. A multi-agent solution to energy management in hybrid renewable energy generation system. *Renew. Energy* **2011**, *36*, 1352–1363. [CrossRef]
115. Dileep, G. A survey on smart grid technologies and applications. *Renew. Energy* **2020**, *146*, 2589–2625. [CrossRef]
116. Rivas, A.E.L.; Abrão, T. Faults in smart grid systems: Monitoring, detection and classification. *Electr. Power Syst. Res.* **2020**, *189*, 106602. [CrossRef]
117. Tan, W.-S.; Shaaban, M. Dual-timescale generation scheduling with nondeterministic flexiramp including demand response and energy storage. *Electr. Power Syst. Res.* **2020**, *189*, 106821. [CrossRef]

118. Milchram, C.; Künneke, R.; Doorn, N.; van de Kaa, G.; Hillerbrand, R. Designing for justice in electricity systems: A comparison of smart grid experiments in the Netherlands. *Energy Policy* **2020**, *147*, 111720. [CrossRef]
119. Rohde, F.; Hielscher, S. Smart grids and institutional change: Emerging contestations between organisations over smart energy transitions. *Energy Res. Soc. Sci.* **2021**, *74*, 101974. [CrossRef]
120. Feng, C.; Wang, Y.; Chen, Q.; Ding, Y.; Strbac, G.; Kang, C. Smart grid encounters edge computing: Opportunities and applications. *Adv. Appl. Energy* **2021**, *1*, 100006. [CrossRef]
121. Pala, D. *ICT Standards for Smart Grids: IEC 61850, CIM and Their Tmplementation in the ERIGrid Project*; ERIGrid: Vienna, Austria, 2018.
122. Umeozor, E.C.; Trifkovic, M. Operational scheduling of microgrids via parametric programming. *Appl. Energy* **2016**, *180*, 672–681. [CrossRef]
123. Lv, T.; Ai, Q. Interactive energy management of networked microgrids-based active distribution system considering large-scale integration of renewable energy resources. *Appl. Energy* **2016**, *163*, 408–422. [CrossRef]
124. Gu, W.; Wu, Z.; Bo, R.; Liu, W.; Zhou, G.; Chen, W.; Wu, Z.Z. Modeling, planning and optimal energy management of combined cooling, heating and power microgrid: A review. *Int. J. Electr. Power Energy Syst.* **2014**, *54*, 26–37. [CrossRef]
125. Zhang, J.; Cheng, H.; Wang, C. Technical and economic impacts of active management on distribution network. *Int. J. Electr. Power Energy Syst.* **2009**, *31*, 130–138. [CrossRef]
126. Mohamed, A.; Mohammed, O. Real-time energy management scheme for hybrid renewable energy systems in smart grid applications. *Electr. Power Syst. Res.* **2013**, *96*, 133–143. [CrossRef]
127. Mah, D.N.-Y.; Wu, Y.-Y.; Ip, J.C.-M.; Hills, P.R. The role of the state in sustainable energy transitions: A case study of large smart grid demonstration projects in Japan. *Energy Policy* **2013**, *63*, 726–737. [CrossRef]
128. Kellerer, E.; Steinke, F. Scalable Economic Dispatch for Smart Distribution Networks. *IEEE Trans. Power Syst.* **2015**, *30*, 1739–1746. [CrossRef]
129. Lamadrid, A.J.; Shawhan, D.L.; Nchez, C.E.M.-S.Á.; Zimmerman, R.D.; Zhu, Y.; Tylavsky, D.J.; Kindle, A.G.; Dar, Z. Stochastically Optimized, Carbon-Reducing Dispatch of Storage, Generation, and Loads. *IEEE Trans. Power Syst.* **2015**, *30*, 1064–1075. [CrossRef]
130. Niknam, T.; Kavousifard, A.; Tabatabaei, S.; Aghaei, J. Optimal operation management of fuel cell/wind/photovoltaic power sources connected to distribution networks. *J. Power Sources* **2011**, *196*, 8881–8896. [CrossRef]
131. Elamine, D.O.; Nfaoui, E.H.; Jaouad, B. Multi-agent system based on fuzzy control and prediction using NN for smart microgrid energy management. In Proceedings of the 2015 Intelligent Systems and Computer Vision (ISCV), Fez, Morocco, 25–26 March 2015; IEEE: Piscataway, NJ, USA, 2015; pp. 1–6.
132. Phuangpornpitak, N.; Tia, S. Opportunities and Challenges of Integrating Renewable Energy in Smart Grid System. *Energy Procedia* **2013**, *34*, 282–290. [CrossRef]

Article

Co-Simulation Framework for Optimal Allocation and Power Management of DGs in Power Distribution Networks Based on Computational Intelligence Techniques

Marinko Barukčić *,†, Toni Varga †, Vedrana Jerković Štil † and Tin Benšić †

Faculty of Electrical Engineering, Computer Science and Information Technology Osijek, Josip Juraj Strossmayer University of Osijek, 31000 Osijek, Croatia; toni.varga@ferit.hr (T.V.); vedrana.jerkovic@ferit.hr (V.J.Š.); tin.bensic@ferit.hr (T.B.)

* Correspondence: marinko.barukcic@ferit.hr; Tel.: +385-31-224-6098

† These authors contributed equally to this work.

Abstract: The paper researches the impact of the input data resolution on the solution of optimal allocation and power management of controllable and non-controllable renewable energy sources distributed generation in the distribution power system. Computational intelligence techniques and co-simulation approach are used, aiming at more realistic system modeling and solving the complex optimization problem. The optimization problem considers the optimal allocation of all distributed generations and the optimal power control of controllable distributed generations. The co-simulation setup employs a tool for power system analysis and a metaheuristic optimizer to solve the optimization problem. Three different resolutions of input data (generation and load profiles) are used: hourly, daily, and monthly averages over one year. An artificial neural network is used to estimate the optimal output of controllable distributed generations and thus significantly decrease the dimensionality of the optimization problem. The proposed procedure is applied on a 13 node test feeder proposed by the Institute of Electrical and Electronics Engineers. The obtained results show a huge impact of the input data resolution on the optimal allocation of distributed generations. Applying the proposed approach, the energy losses are decreased by over 50–70% by the optimal allocation and control of distributed generations depending on the tested network.

Keywords: co-simulation; computational intelligence techniques; distributed generation; optimal allocation and control

Citation: Barukčić, M.; Varga, T.; Jerković Štil, V.; Benšić, T. Co-Simulation Framework for Optimal Allocation and Power Management of DGs in Power Distribution Networks Based on a Computational Intelligence Techniques. *Electronics* **2021**, *10*, 1648. https://doi.org/10.3390/electronics10141648

Academic Editor: Cheng Siong Chin

Received: 24 May 2021
Accepted: 8 July 2021
Published: 10 July 2021

1. Introduction

The optimal allocation and power control of distributed generations (DGs) in electrical power systems has been the focus of researchers in recent years, especially in the context of the smart grid concept. In the project planning phase of DGs installation in the power network, the optimal allocation should determine the system nodes/buses and sizes/powers of the DGs. After the DGs are installed, in the operational phase of the projects, the optimal control of the controllable DGs outputs is of interest. There are different approaches to solve such optimization problem, considering these two parts simultaneously or separately. Besides this question, there is a research challenge regarding the input data resolution used in the optimization process. Generally speaking, the simultaneous approach and higher input data resolution require more computational effort to solve the problem. As far as modeling of the system in the optimization problem is concerned, there are two approaches used in the research. One is based on the usage of an analytical model (a system of equations) of the power network, and the other uses the simulation tool to calculate the objective and constraint functions of the optimization problem. The analytical approach usually means more approximations, and neglecting will be included in the model of the system, decreasing the realistic representation of the system. On the other hand, nowadays, the simulation tool for power system analysis ensures less neglecting, resulting in more realistic modeling of the system. The results obtained using more realistic model in the optimization process are more reliable for practical implementation. These two research issues in this topic are the research subject of this paper. Besides the single or multi-objective approach, the different objective functions as well as the problem constraints of the optimization problem are used in research studies of optimal allocation and the control of DGs. A brief overview of the literature on the topic concerning the above-mentioned issues and research challenges is presented below in the section.

In [1] the authors used the fuzzy system to aggregate the multi-objective problem into single-objective optimization, considering the economic and environmental objective functions and technical constraints. The objective functions were calculated from analytical expressions, and node sensitivity analysis (gradient-based optimization) was used to find the optimal DG allocations. The optimization assumed constant (static) load powers in the power network. The load powers dependent on a bus voltage were considered in [2] to find the optimal allocation of DGs. However, the nominal loads were taken to be constant—not changing in time. The single-objective objective function value and technical constraints were used and solved by the metaheuristic optimization method (Harmony Search Algorithm—HAS). The calculation of the objective function values was coded in the programming environment in which the optimization method was implemented. The multiobjective optimization problem of optimal allocation and control of Battery Energy Storage Systems (BESS) is presented in [3]. Particle Swarm Optimization (PSO), which belongs to the metaheuristic techniques, was used to solve the problem with two objectives: power losses and total power of installed BESS. Optimal allocation was performed with constant nominal load values. After the optimal BESS allocations were found, the optimal charging/discharging control of BESSs with the hourly resolution was determined, using the proposed analytical method. In [4], the single optimization problem considering the DG penetration level as objective and total harmonic distortion (THD) as the problem constraint were investigated again with static load values. The problem was solved by metaheuristic optimization techniques, namely Genetic Algorithm (GA) and Differential Evolution (DE). In addition, in [3,4], objective function values were coded together with the optimization method. In [5], the multiobjective optimization of the optimal power generation of Virtual Power Plant (VPP, including DG and EV (Electrical Vehicle) charging stations) power generation was solved, using the PSO method. The daily load shape with the hourly resolution was used in the simulations. Two-objective optimization, considering the operational cost and which are the pollutants emission, was presented in [6]. The problem was solved using the Ant Lion Optimizer (ALO) metaheuristic method, and variable loads on a daily level with hourly resolution were used. The optimization procedure, as well as the problem, were coded together inside the same programming tool. The single objective optimization dealing with losses minimization and DG penetration level maximization was presented in [7] to find optimal allocation of DGs. The daily profile of load, Photovoltaic (PV), and Wind (W) generations DG units with minute resolutions for 12 typical days were used in the research. The objective function values were calculated by coding the problem in a programming tool, and it is not clear what optimization method was used here. The single objective optimization problem of optimal DG allocation was solved by applying the metaheuristic optimizer, Grey Wolf Optimizer (GWO) in [8]. The optimization was performed for a constant load value, and methods for power flow calculations and optimization itself were coded with the same computational tool. In [9], single-objective optimization, considering the optimal allocation of DG for reactive power control, was solved by using GA. The co-simulation approach was used here, employing the external power system simulator to calculate the objective function value. As in most literature considering the topic, the optimization was performed for constant load value. The constant loads were considered in [10] during single-objective optimization, aiming to find the optimal DG allocations with power losses minimization. The objective function values were calculated based on the backward/forward sweep power flow method which was coded alongside the optimization method in the same programming environment. Different metaheuristic techniques, HAS, Artificial Bees Colony (ABC), and PSO, were used to solve the optimization problem. The research [11] presented a co-simulation approach to solve the optimal DG allocation problem considering one objective function and constant loads. The single objective optimization of DG allocation was solved by using the Sensitivity Analysis (SA) method in [12]. The optimization method and calculation of the objective function value were implemented in the same programming tool. The yearly profile (at hourly data resolution) of wind DG production was considered, but it is not clear if the load profile was used too or if the loads were assumed to be fixed. In [13] the multiobjective optimal DG allocation problem was solved by using the metaheuristic optimization method Ant Lion Optimizer (ALO). The optimal allocation of DGs, BESS, and reactive power control devices was found, considering constant loads. The paper [14] dealt with optimal DG and BESS allocations, solving the problem by using mixed-integer conic programming (MICP) as an optimization technique. The optimal allocation problem was defined in a form of single-objective optimization. The optimization problem was modeled in the specific modeling tool, and an existing external optimization tool was interfaced to the model to find the optimal solution. The optimization problem considered load and DG production profiles with hourly resolution at the yearly level. However, the clustering technique was used to generate typical 48 profile patterns to decrease the dimensionality of the problem. The authors in [15] used PSO to solve the single optimization problem of optimal DG allocation. The optimization process considered constant load and power flow calculation, as well as the optimization algorithm, and was coded in the same programming tool. In [16], the hybridization of two metaheuristic methods, PSO-SFL (Shuffled Frog Leap), was used to solve the single optimization problem of optimal DG allocation. The constant load values were considered and power flow calculations were implemented in the programming tool used for the optimization method performing. The research presented in [17] solved the optimal allocation of DGs, Shunt Capacitors (SC), and Electrical Vehicle (EV) charging stations by using the Grasshopper Optimization Algorithm (GOA), which belongs to the class of metaheuristic methods. The proposed procedure solved the single-objective optimization problem (with four objectives aggregated into one objective function) in two separated steps considering the optimization of DG and SC allocations separately for the optimization of the EV charging station allocation. The constant load values were used during the solving of the optimization problem, and the impact of the load and DG production changes was investigated once the optimal DG and SC allocations were determined. The power flow analysis was performed in the same programming tool, which was used for the optimization method implementation. A hybrid metaheuristic method, GA-PSO, was applied in [18] to solve a single-objective problem with the aggregated objective function consisting of three parts. The load values were assumed to be constant, and the power flow calculation was implemented in the optimization method objective function value calculation. The authors in [19] applied GA optimization to solve the single optimization problem of optimal DG and BESS allocations, considering the daily load shapes with hourly resolution. The objective function calculation was coded inside the optimization procedure in the PYTHON programming environment. In [20], the single-objective optimization problem of optimal DG allocation was solved by the DE optimization algorithm. The optimal DG allocation, as well as the DG power factor, were decision variables of the optimization problem. The load and DG production profiles were considered during the optimization. The daily profiles with hourly resolution were used, and these profiles were obtained by averaging values from the seasonal profiles. The DIgSILENT simulation tool was employed for the power flow calculation based on which the objective function value was calculated. Although not clearly stated, if the optimization method was implemented inside (built-in) the DIgSILENT or in some external programming tool, it seems that the co-simulation approach was used here.

Based on the above-given brief overview of research studies dealing with optimal DG allocation, the description of the reviewed literature can be summarized as follows:

- Most of the literature considers constant/static load and DG production in the distribution network.
- If a variable load/DG production is considered, then the changes are usually observed on a daily level with hourly averaged values/resolution.
- Much of the literature used an approach in which both the power flow calculations (which is the base for objective function values calculation) and an optimization algorithm were coded in a programming environment.
- The optimal allocation of DGs and optimal control (power management/dispatch) problems were solved separately.

The above literature review yields three main issues considering the optimal allocation of DGs:

- The influence of choosing constant or variable load and generation values on DG optimal allocation results.
- The choice of a proper approach for applying the optimization solver and objective calculations—one simulation tool or co-simulation tools.
- An approach to solve the optimization of the allocation and power management of DGs—separately or simultaneously, the optimization of the allocation and power management.

These previously mentioned are detected as open research questions, challenges, and gaps in the topic of optimal DG allocation. The presented research aims to decrease the research gaps and make a contribution to the topic through the following:

- Propose the framework for the co-simulation approach, using in the optimization of DG allocation the power distribution network with the aim of more realistic distribution system modeling.
- Propose (inside the co-simulation framework) the application of the computational intelligence techniques to decrease the dimensionality of the optimization problem and handle uncertainties in the power system.
- Simultaneously perform allocation optimization and DG power management.

The existing literature includes the above-listed aspects (some or all). The optimal DG allocation problem considered in this paper can be summarized as follows:

- Literature [6,7] dealt with variable load values.
- Literature [9,11,14] dealt with the co-simulation approach.
- Literature [20] dealt with variable load and generation profiles as well as the co-simulation approach.

The approach of simultaneously considering the location, size, load, and DG production profiles during the optimization used in this research study is similar and with similar aims as that presented in [20], which was one of the inspirations for this research.

The rest of the paper is organized as follow: the optimization problem formulation is given in Section 2; Section 3 describes the applied research methodology and proposed framework, including a brief overview of the used specific simulation tools; the results of different scenarios aiming to validate the proposed framework applied on a test power distribution network are presented in Section 4; the discussion about the obtained results concerning the stated research question regarding the input data resolution impact is given in Section 5; and at the end, some general remarks/conclusions are presented in Section 6.

2. The Optimization Problem of the DGs Allocation

The optimization problem is defined to address the research gaps mentioned above, in the previous section. In the literature, different objectives are considered; here, two objective functions that are important from the point of view of the power distribution system operator and the owners of the DGs are used. As one of the main interests of the power distribution system operator is decreasing the losses, the first objective function used in optimization is the active energy losses W_{loss}. Generating as much energy from DGs as possible is the main interest of the DG owners to shorten the investment payback period and increase the profit. This leads to increasing the penetration level of DGs in the power distribution system. Such an objective is formulated here in the form of the total exchange of the total apparent energy W_S in the power coupling point of the distribution network on the upstream network. Because these two objective functions are conflicted, the multiobjective (two-objective) optimization approach is used in the research. The technical constraints regarding the nodal voltage range and line current limits as well as the box constraints of the decision variable ranges are applied in the optimization problem.

The mathematical notation of the multiobjective optimization problem, including the previously described objective functions and constraints, is as follows:

$$\begin{array}{l} F = [W_{loss}(x), W_S(x)] \rightarrow \min \\ \text{subject to box constraints:} x \in \{x_{lb}, x_{ub}\} \\ \text{subject to inequality constraints:} \\ V_{min} \leq V_{i,e} \leq V_{max}, I_{k,e} \leq I_{k,max} \leq \epsilon \\ \text{with decision variables vector: } x, \end{array} \tag{1}$$

with the following notations: F—two objective problem function consisting of W_{loss}, yearly energy losses in the network, and W_S, yearly exchange of the apparent energy between a network with DG and upstream system; x_{lb}—lower bounds of the decision variable values; x_{ub}—upper bounds of the decision variable values; V_{min}—lower bounds of the nodal voltage value; V_{max}—upper bounds of the nodal voltage value; $V_{i,e}$—calculated nodal voltage in the i-th network node; $I_{k,max}$—maximum allowed currents in the k-th network line; and $I_{k,e}$—calculated current in the k-th network line.

The objective functions represent energies over the timespan and for N time segments t_i are calculated from active power losses (P_{loss}), active (P_{exc}) and reactive (Q_{exc}) powers exchanged with the upstream network as follows:

$$\begin{array}{c} W_{loss} = \sum_{i=1}^{N} P_{loss,i} \cdot t_i \\ W_S = \sqrt{\left(\sum_{i=1}^{N} P_{\text{exc, i}} \cdot t_i\right)^2 + \left(\sum_{i=1}^{N} Q_{\text{exc, i}} \cdot t_i\right)^2} \end{array} \tag{2}$$

Optimization Problem (1) has a two-objective function that consists of two objectives: yearly energy losses in the network (W_{loss}) and yearly exchange of the apparent energy between a network with DG and upstream system (W_S). The problem constraints in (1) are related to the decision variable ranges (box constraints) and the network operational constraints. The box constraints represent ranges of the decision variable values. The operational constraints are related to the standardized nodal voltage ranges ($V_{min} - V_{max}$; usually nodal voltage limits are given in range ±5% or ±10% depending on the relevant standard) and rated currents of the network lines ($I_{k,max}$).

Optimization Problem (1) is solved by using the Pareto dominance definitions resulting from the solution set known as the Pareto set [21]. Except for the multiobjective optimization, the two single objective optimizations considering each of the objective functions individually are performed also to check if the multiobjective approach can find reliable edges of the Pareto set. The three different problem setups are used in the research, each of them resulting in a different number of decision variables. This being the case, the contents of the decision variables vector is detailed later in the text in Section 3, describing the proposed procedures. The co-simulation setup of the distribution power system simulation and tools of the computational intelligence methods are used to solve the black-box optimization problem model.

The simulation tool for electrical power system simulation is used to calculate the objective function values. Both DGs types with controllable and non-controllable primary energy sources are considered in the distribution system model. The non-controllable DGs used in the research are Photovoltaic (PV) and wind (WD) plants; as controllable DGs, Biogas (BG) plants are used in the distribution power network. For non-controllable energy sources, the production profiles (the DGs outputs) are involved in the model; for the controllable source, the DGs outputs are subject to optimization. In addition, the load profiles (costumers load shapes) are considered in the optimization. The base case in the simulations supposes DG production and load profiles on a yearly level with hourly resolution, i.e., with 8760 data.

2.1. *A Brief Overview of the Used Tools in Co-Simulation*

The optimization tool applying the metaheuristic optimization technique MIDACO solver (Mixed Integer Distributed Ant Colony Optimization) [22] is used to solve the optimization problem. The advantage of this tool is its applicability in a general case of complex optimization problems, including continuous (linear (LP) and nonlinear (NLP)), integer (discrete) (IP), and mixed-integer (MINLP) problems. This tool is based on the Ant Colony Optimization (ACO) [23] which belongs to the metaheuristic methods. The MIDACO can handle single as well as multiobjective optimization.

The artificial neural network (ANN) is applied in the optimization problem, aiming to decrease the number of the decision variables, reducing the dimensionality of the problem. The Multilayer Perceptron (MPL) ANN with one hidden layer is used here. The ANN is modeled by usage of the TensorFlow tool [24] by applying the Keras API [25].

The OpenDSS simulation tool [26] is used in the study with the purpose of solving the power flow calculations of the model of the distribution network and obtain the objective function values needed for the performing optimization procedure. The usage of the power network simulation tool enables including more details in the network model, decreasing the approximations and neglecting in the model, compared to an analytically defined objective function. This ensures more realistic power distribution network modeling, resulting in more reliable results obtained through the optimization.

As mentioned before, all three of these computational tools are implemented in the Python programming environment and employed in the co-simulation setup to solve the black-box optimization problem.

3. Co-simulation Framework for DG Allocation and Power Management Optimization

The research methodology used in the study is based on the implementation of different scenarios for optimization and the investigation of the input data resolution impact on the solution of the optimization problem. The optimization scenarios are related to variable or fixed power factors of each DG and directly determine the number of decision variables in the optimization problem. The three cases described in Table 1 are used as optimization scenarios.

Table 1. The different optimization scenarios.

Scenario	Uncontrollable DG Output	Uncontrollable DG Power Factor	Controllable DG Output	Controllable DG Power Factor
Opt 1	by energy source profile	fixed	by ANN	fixed
Opt 2	by energy source profile	fixed	by ANN	by ANN
Opt 3	by energy source profile	by ANN	by ANN	by ANN

As can be seen in Table 1, the values of the DG power factor can be directly included as a decision variable, and in that case, the fixed value of the power factor is optimized and does not change with DG power output changes. The other scenario investigates application of the ANN to estimate the optimal DG power factor and in this case, the DG power factor changes over time. For all scenarios, the output power of controllable DG is determined by the ANN. These three basic cases will result in different types and numbers of decision variables in the optimization problem. In Table 2, an overview of the decision variables that occur in the described optimization scenarios is presented.

Table 2. The optimization decision variables for different scenarios.

Scenario	DG Locations	DG Output	ANN Weights and Biases for Controllable DG Power Values	ANN Weights and Biases for Controllable DG Power Factor Values	ANN Weights and Biases for Uncontrollable DG Power Factor Values
Opt 1	✓	✓	✓	x	x
Opt 2	✓	✓	✓	✓	x
Opt 3	✓	✓	✓	✓	✓

In all scenarios, the consumers' load shapes and energy source profiles are used to model the time variability of the primary energy sources and consumption.

In Figures 1–3, the proposed frameworks according to the previously described scenarios are shown. Application of the ANN to estimate the optimal DG power output and DG power factor values is one of the main contributions of this research. The purpose of the ANN is to decrease the number of decision variables in the optimization problem. Without the proposed usage of the ANN for each controllable (in some scenarios, also for uncontrollable (Table 2)) source, the number of decision variables will be equal to the input data resolution, e.g., for yearly input data with the hourly resolution it will be an additional 8760 decision variables per DG for scenarios Opt 1, i.e., twice for scenario Opt 2 or four times for scenario Opt 3 (Table 2).

Based on the scenarios overview given in Table 1 and Figures 1–3, the similarity and differences between the optimization models Opt 1–Opt 3 can be summarized as follows. For all optimization scenarios, uncontrollable DG outputs are considered variable in time and defined by energy source profiles. The controllable DG output is managed by the ANN in all scenarios. The differences among the scenarios are related to the power factor variability for controllable and uncontrollable DGs. In Opt 1, the fixed DGs power factors of both controllable and uncontrollable DGs are optimized. The fixed and ANN managed power factors of the controllable and uncontrollable DGs, respectively, are optimized in scenario Opt 2. In Opt 3, modeling the power factors of all DGs (controllable and uncontrollable) is managed by ANN.

The details about the ANN inputs and outputs are given in Figure 4, and the ANN parameters (weights and biases) are optimized by the MIDACO solver simultaneously with DGs allocation optimization. The simple Multilayer Perceptron (MLP) ANN consists of one hidden and output layer. Because the usage of the ANN has the purpose of significantly decreasing the number of decision variables, this simple ANN configuration is implemented in the framework. The purpose of ANN is to significantly decrease the number of problem decision variables. If the ANN is not used, the number of decision variables would be increased for the number of the input data (depending on the data resolution, e.g., in case of yearly data with hourly resolution the optimization problem would have 8760 data only for the DG power management). The ANN takes as inputs the *i*-th data from each load shape and each uncontrollable DG output profile. Training of the ANN was performed in this way (without application of standard built-in ANN optimizer presents in the used ANN tool) because the training process, in this case, is slightly different than ordinary. Usually, when ANN is used for regression purposes, the difference between the target and ANN estimated values is objective in the ANN optimization process (training). However, this is not the case here; the ANN target outputs are not known in advance, as they need to be determined during the optimization of the whole problem defined by (1). Such configuration of the proposed framework and purpose of the ANN prevents performing common ANN training procedures.

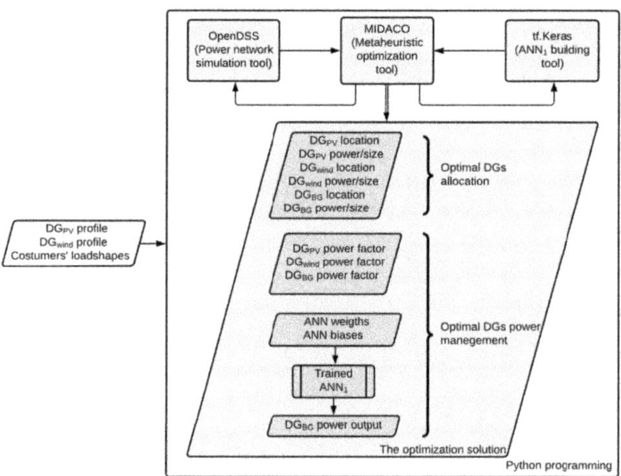

Figure 1. The framework used for Opt 1 optimization scenarios.

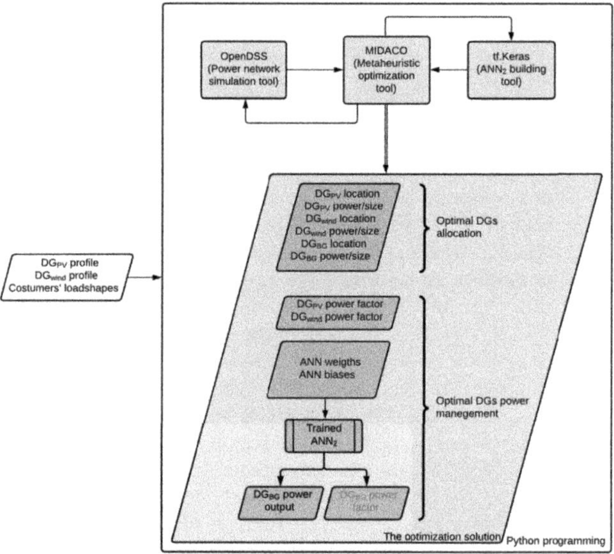

Figure 2. The framework used for Opt 2 optimization scenarios.

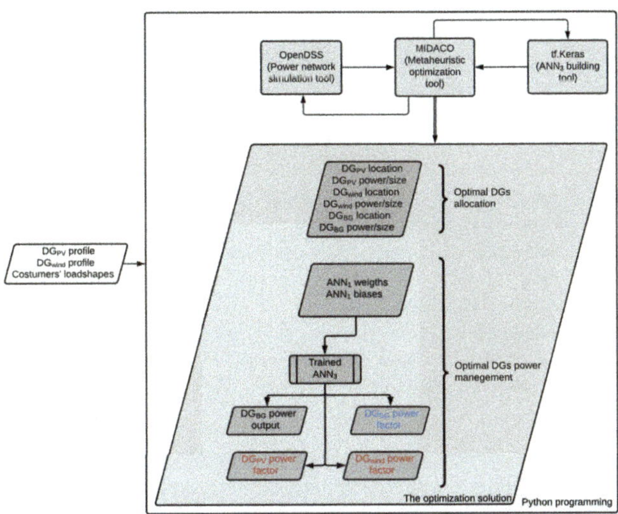

Figure 3. The framework used for Opt 3 optimization scenarios.

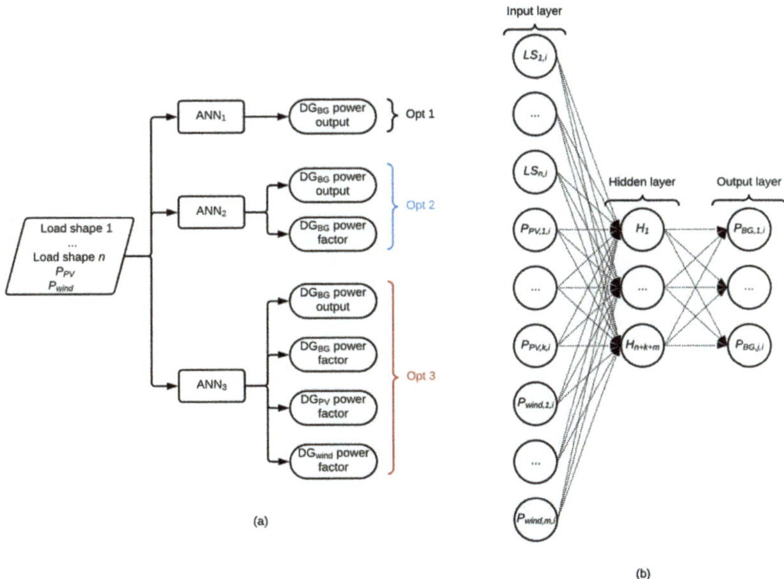

Figure 4. (**a**) Overview of data types used by the ANN; (**b**) schematic structure of the ANN.

The two different workflows are applied to investigate the possible improvement of the solution quality for scenarios Opt 2 and Opt 3 (Figure 5). The difference between the workflows is about the initial solution that the optimization algorithm starts with. In both workflows (WF1 and WF2), the initial solution is randomly generated in Opt 1 optimization scenarios. In workflow W1 the initial solution is also randomly chosen for scenarios Opt 2 and Opt 3, while in workflow WF2 the starting optimization point for Opt 2 and Opt 3 is the solution of solved Opt 1. In the case of WF2 workflow, the locations, and sizes of all DGs and trained ANN for controlling the power output of controllable DG are the solution from Opt 1, and only parameters of the ANN are used to optimize the DG power factors. The purpose of the WF2 workflow is to check if the optimal solution obtained by solving Opt 1 can be additionally improved by managing the DGs power factor.

The above-described optimizations and workflows are performed with input data on yearly level with hourly resolution, i.e., with 8760 input data modeling changes of primary energy source intensity and loads over a year. After the research done according to the workflows, the final workflow and optimization model will be proposed based on a comparison of the obtained solutions for different optimization scenarios. In the continuation of the proposed procedure, the previously chosen optimization scenario and workflow will be repeated with the different decreased resolutions of input data to research the impact of data resolution on the optimization problem solution. The two decreased input data resolutions with 12 and 365 input data are used here. These input data are obtained by averaging the basic input data (hourly resolution, 8760 input data) on monthly (12 input data) and daily (365 input data) levels. The schematic overview of this part of the research methodology is given in Figure 6.

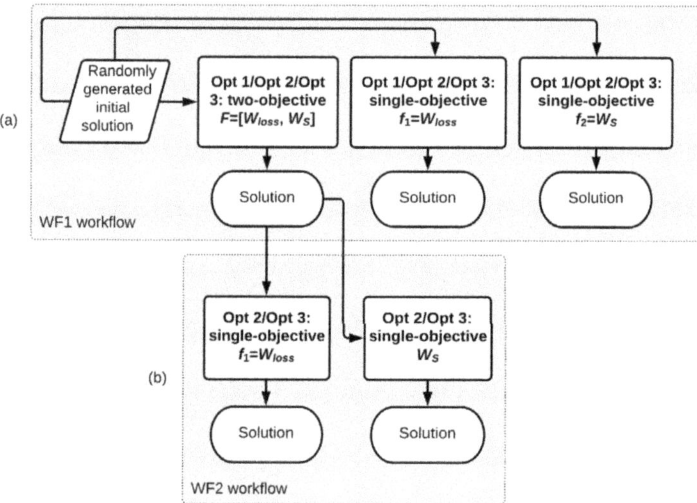

Figure 5. The workflows used in the research study, (**a**) workflow uses randomly generated input data for single and multi-objective optimizations; (**b**) workflow uses solution of the multi-objective optimization as input to single-objective optimizations.

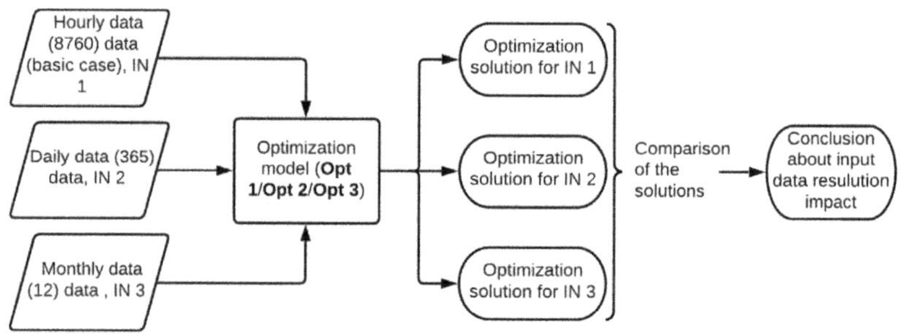

Figure 6. Application of input data with different resolutions in the proposed procedure.

At the end of the research methodology used in this study, the chosen optimization procedure optimized with hourly input data was applied, using input data with a resolution of 15 min (four times higher resolution than one used in optimization procedure) steps, i.e., with 35,040 input data. The load shapes are one of the inputs into the ANN (Figure 4). Usually, the load shapes are forecasted, based on the historical consumption data, and consequently include more or less uncertainty. Due to this uncertainty, the later application of the load shape can give a load value that is different from the real load value. This procedure part is performed with two scenarios: one supposing an unknown real load value at the moment and the other with a known specific load value at the specific time step. If the real load value is not known/measured, the load value is estimated according to the load shapes used in the optimization process. The previously described is visualized in Figure 7. The purpose of this is to investigate the robustness of the proposed framework. In addition, with the development of the smart grid concept, it is expected more and more the usage of smart energy meters at the costumers' point of common coupling. These conditions allow obtaining the load values with more accuracy as input data during the implementation of the proposed framework.

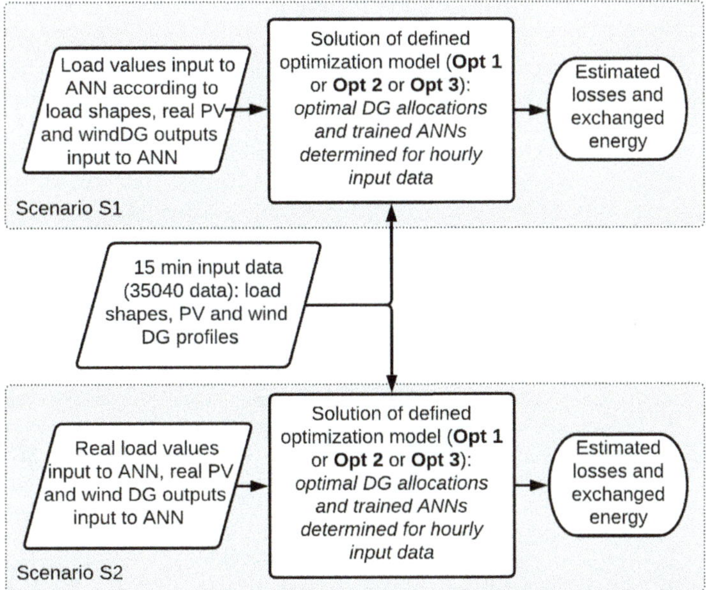

Figure 7. Scenarios for the procedure application for 15 min resolution of input data.

4. Application of the Framework on Test Distribution Power Systems—Case Study

The very well-known and often used IEEE 13 node IEEE 37 node test feeders [27] are used as a case study to demonstrate the application of the proposed framework for optimal DG allocation and power management. These distribution power systems are chosen because they represent the most general examples of the power networks. Some features of the used power networks are an unbalanced system (one, two, and three-phase lines and loads), different load models (constant power, constant impedance, constant current ...), voltage regulators, load in wye (star), and delta connection configurations and two voltage levels (4.16 kV and 0.48 kV). All details about the used test systems can be found in [27]. The next modification in the original IEEE 13 node and IEEE 37 node bus test feeders is made to adapt the systems for performed research: the capacitor banks are turned off, the taps of the voltage regulators are reset to the middle position before the simulation for each possible solution in the optimization is started, and three different load shapes are assigned to the consumers according to Table 3. The proposed optimization procedure is performed on a desktop PC with Intel i7-10700 CPU 2.90 GHz, 8 Cores, RAM 16 GB. The versions of the used software are PYTHON 3.8, MIDACO 5.0, OpenDSS 9.1.3.2, and Tensorflow 2.3.1. Data about the optimization parameters used in the optimizations are as follows: number of function evaluation in MIDACO is 30,000, number of ants and kernels in MIDACO are 250 and 15, respectively. The computational times are 1.1 and 6 s per iteration for IEEE 13 and 37 bus test networks, respectively. The total computational times for these two tested networks are about 9 and 22 h for IEEE 13 and 37 bus networks, respectively.

Table 3. The load profile of costumers used in the test networks.

Load Profile	Network Bus IEEE 13	Network Bus IEEE 37
LP 01	671, 611, 652, 670	701, 722.3.1, 724, 725, 733–735, 742.2.3
LP 02	634, 645, 646, 692	712, 713, 714.1.2, 727–729, 736–738, 744
LP 03	675	714.2.3, 718, 720, 722.2.3 730, 732, 740, 741, 742.1.2

4.1. *Input Data Preparation*

The input data regarding the load shapes and PV and wind plant power profiles are obtained by usage of the existing tools for prediction/forecasting load consumption and PV and wind DG production. The computer tool Load Profile Generator (LPG) (https://www.loadprofilegenerator.de/ (accessed on 8 April 2021)) [28] is used to synthesize the three different load profiles used in the simulations. The built-in (LPG) load shape types, namely "H01 in HT 14", "H01 in HT 11", and "H01 in HT 07", are used to generate load profiles LP 01, LP 02, and LP 033, respectively. The normalized load shapes are given in Figure 8a–c by showing example daily profiles in Figure 8e–f.

The online platform (tool) named "Renewables.ninja." (Available online: https://www.renewables.ninja/ (accessed on 8 April 2021) based on research presented in [29,30] is used to produce the output profile of PV and wind DGs. The generated normalized DG production profiles are shown in Figure 9.

The above-described input data, the yearly load, and DG production profiles with the hourly resolution are generated as basic input data cases. The input data with monthly and daily resolutions used as input into part of conducted research study shown in Figure 6 are generated from the base input data by averaging data on monthly and daily levels, respectively. These input data with the decreased resolution are presented in Figures 10–13.

The input data used at the end of the conducted research shown in Figure 7 are produced by adding randomly generated noise (X according to uniform distribution $\mathcal{U}$) to the base data for each time step (1 h) four times. The repeated random number generation four times for each hour produces load and DG output profiles with a resolution of 15 min (1 h = 4 × 15 min). The noise range is set in the range $\pm z\%$ of the hourly value. The mathematical formulation of the 15 min resolution data is the following:

$$LF_{15min} = LF_h + X \sim \mathcal{U}(-z \cdot LF_h,\ z \cdot LF_h) \tag{3}$$

In this case, it is important to highlight that hourly data are not averages of 15 min data. Due to higher data resolution, the 15 min resolution data are not visualized for the whole year, but only for the first day in a year (in comparison with basic input data (hourly resolution)) as can be seen in Figures 14 and 15 (for example, for ±20% uncertainty used in the 15 min data generation).

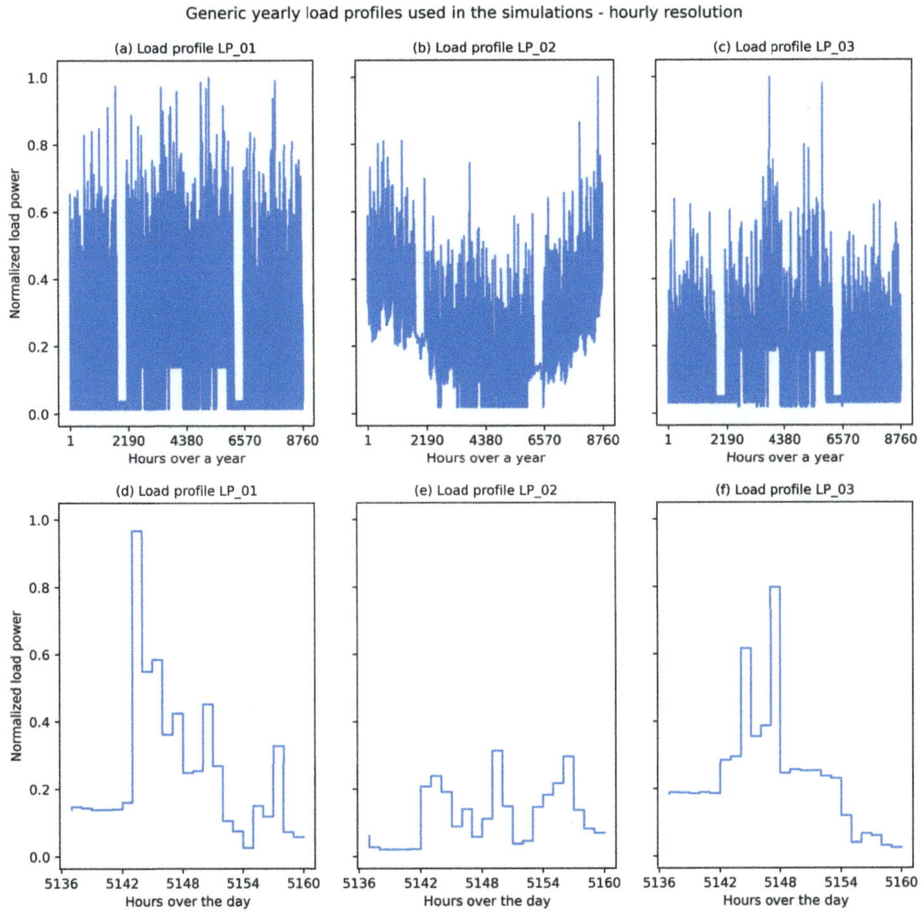

Figure 8. Input data—load shapes: (**a**) LP 01-year; (**b**) LP 02-year; (**c**) LP 03-year; (**d**) LP 01-day; (**e**) LP 02-day; (**f**) LP 03-day.

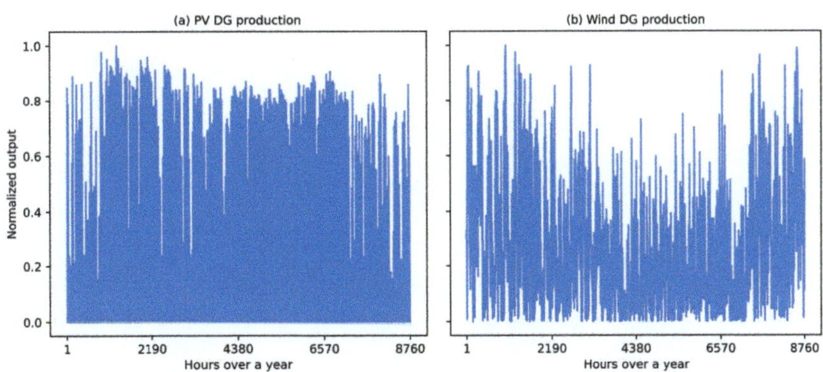

Figure 9. Input data—load shapes: (**a**) PV plant output; (**b**) wind plant output.

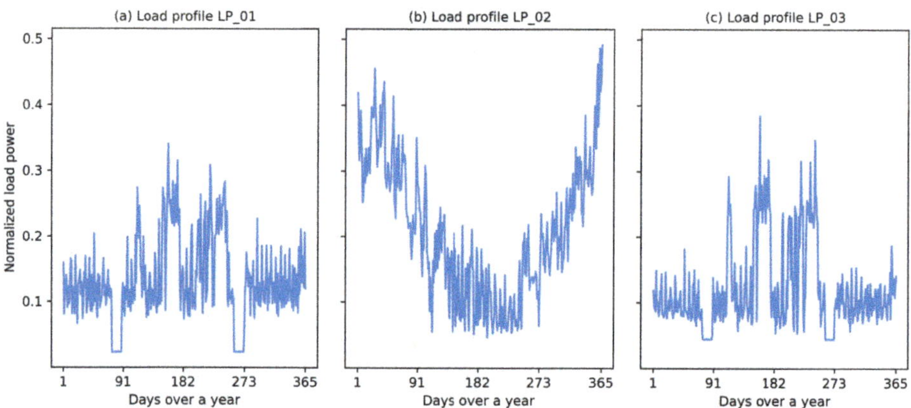

Figure 10. Load shapes with daily resolution: (**a**) LP 01; (**b**) LP 02; (**c**) LP 03.

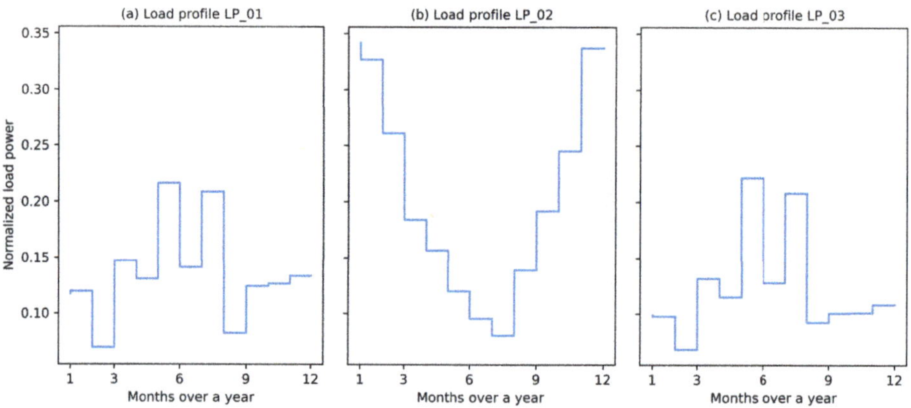

Figure 11. Load shapes with monthly resolution: (**a**) LP 01; (**b**) LP 02; (**c**) LP 03.

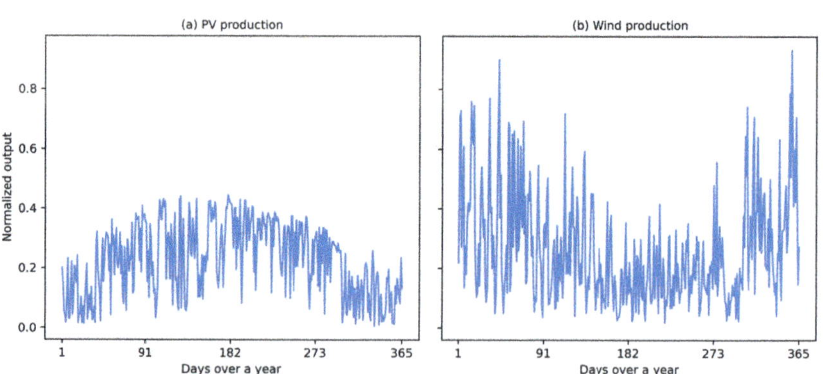

Figure 12. Production profile of DGs with daily resolution: (**a**) PV DG; (**b**) wind DG.

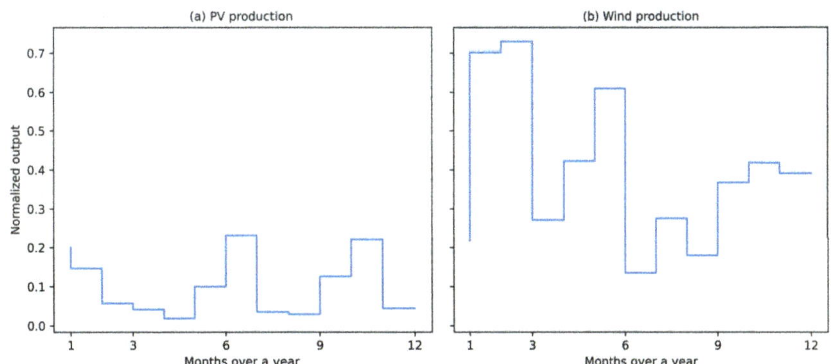

Figure 13. Production profile of DGs with monthly resolution: (**a**) PV DG; (**b**) wind DG.

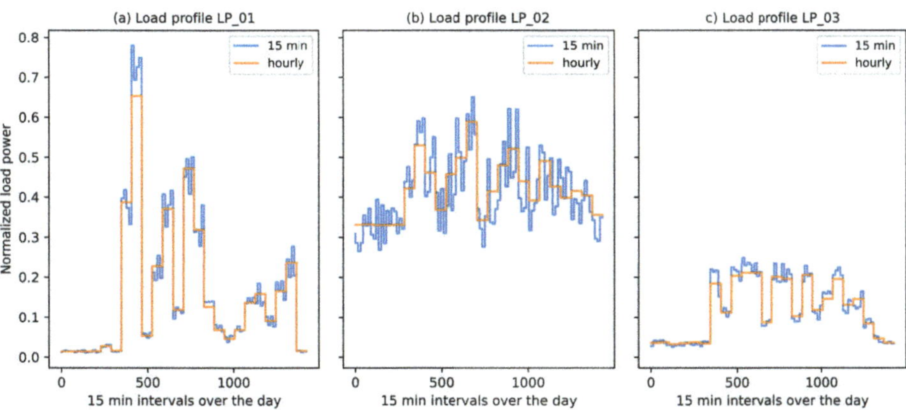

Figure 14. Load profiles with 15 min resolution generated (according to (3)) from hourly resolution (for example, for the first day in a year): (**a**) LP 01; (**b**) LP 02; (**c**) LP 03.

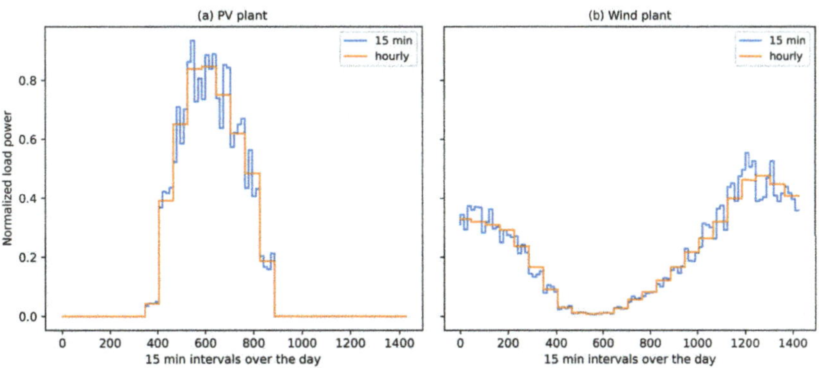

Figure 15. DG output profiles with 15 min resolution generated (according to (3)) from hourly resolution (for example, for the first day in a year): (**a**) PV DG; (**b**) Wind DG.

In Sections 4.2–4.6, the results of the framework implementation are presented.

4.2. Results for Different Optimization Models—Workflow WF1

In this subsection, the results obtained according to workflow W1 (Figure 5a) for the optimization models proposed in Figures 1–3 and Table 1 are presented. Figures 16–18 visualize solutions of optimization models Opt 1 (Figure 1), Opt 2 (Figure 2) and Opt 3 (Figure 3), respectively, showing the Pareto front of the solved optimization Problem (1).

In Tables 4–6, the numerical values for the solutions obtained for all three proposed optimization models (Opt 1, Opt 2 and Opt 3) applied on multi-objective and single objective optimizations are given. For multiobjective optimization, the values for the Pareto edges (for solutions giving the lowest energy losses and energy exchange) are shown in these Tables.

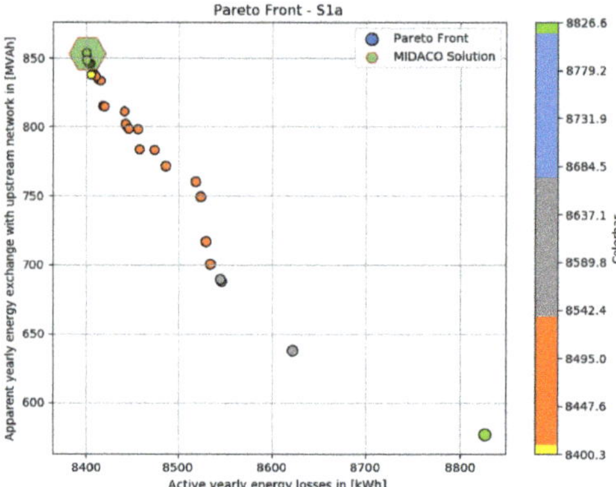

Figure 16. The Pareto front of the solved multi-objective optimization problem for Opt 1 (Figure 1).

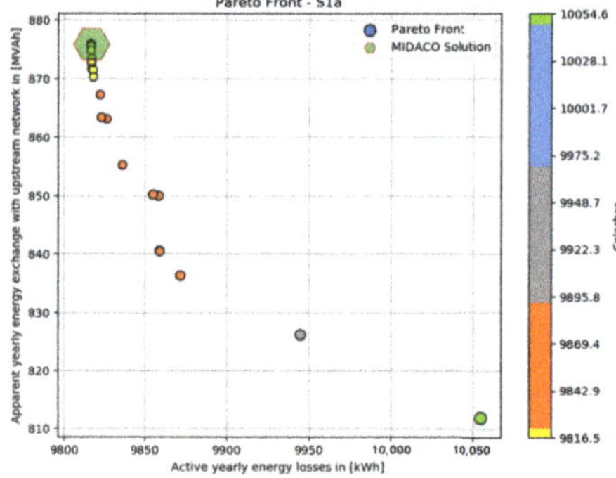

Figure 17. The Pareto front of the solved multi-objective optimization problem for Opt 2 (Figure 2).

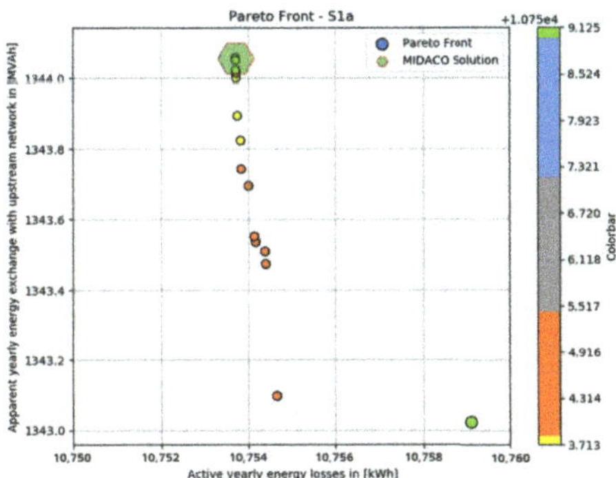

Figure 18. The Pareto front of the solved multi-objective optimization problem for Opt 3 (Figure 3).

Table 4. Results for optimal solutions of Opt 1 optimization model for different objective functions (W1 workflow).

Optimized DG Allocations		W_{loss} in [kWh]	W_S in [MVA]	Decreasing of W_{loss} in [%]	Decreasing of W_S in [%]
Without DGs		39,859	5555.03	-	-
Two-objective	W_{loss} edge	8400	853.45	78.93	84.64
(W_{loss}, W_S)	W_S edge	8826	576.87	77.86	89.62
single objective (W_{loss})		8400	853.45	78.93	84.64
single objective (W_S)		15,979	288.27	59.91	94.81

Table 5. Results for optimal solutions of Opt 2 optimization model for different objective functions (W1 workflow).

Optimized DG Allocations		W_{loss} in [kWh]	W_S in [MVA]	Decreasing of W_{loss} in [%]	Decreasing of W_S in [%]
Without DGs		39,859	5555.03	-	-
two-objective	W_{loss} edge	9816	875.92	75.37	84.23
(W_{loss}, W_S)	W_S edge	10,054	811.86	74.78	85.39
single objective (W_{loss})		9816	875.92	75.37	84.23
single objective (W_S)		16,229	384.02	59.28	93.09

Table 6. Results for optimal solutions of Opt 3 optimization model for different objective functions (W1 workflow).

Optimized DG Allocations		W_{loss} in [kWh]	W_S in [MVA]	Decreasing of W_{loss} in [%]	Decreasing of W_S in [%]
Without DGs		39,859	5555.03	-	-
two-objective	W_{loss} edge	10,754	1344.06	73.02	75.80
(W_{loss}, W_S)	W_S edge	10,759	1343.01	73.00	75.82
single objective (W_{loss})		10,974	1392.83	72.47	74.39
single objective (W_S)		11,503	1089.30	71.14	80.39

4.3. Results for Different Optimization Models—Workflow WF2

The results obtained according to workflow WF2 (Figure 5b) for the optimization models Opt 2 and Opt 3 are presented in this subsection. In WF2, the workflow of the optimal allocations of DGs and trained ANN obtained as a solution of Opt 1 model are used as the initial (starting) solution for the optimization of the Opt 2 and Opt 3 models. In this case, the problem solutions are trained ANNs for controlling DGs power factors. In this workflow, only a single-objective optimization approach is applied based on the results obtained in the previous subsection. As can be seen from the results obtained in workflow WF1, the single-objective optimization finds the Pareto front edges of better quality. The results obtained according to workflow WF2 are shown in Tables 7 and 8 for optimization models Opt 2 and Opt 3, respectively.

Table 7. Results for optimal solutions of Opt 2 optimization model for different objective functions (W2 workflow).

Optimized DG Allocations	W_{loss} in [kWh]	W_S in [MVA]	Decreasing of W_{loss} in [%]	Decreasing of W_S in [%]
Without DGs	39,859	5555.03	-	-
single objective (W_{loss})	8391	851.89	92.94	84.64
single objective (W_S)	15,973	271.14	59.93	95.12

Table 8. Results for optimal solutions of Opt 3 optimization model for different objective functions (W2 workflow).

Optimized DG Allocations	W_{loss} in [kWh]	W_S in [MVA]	Decreasing of W_{loss} in [%]	Decreasing of W_S in [%]
Without DGs	39,859	5555.03	-	-
single objective (W_{loss})	12,387	1843.40	69.92	66.82
single objective (W_S)	20,833	2369.38	47.73	57.35

4.4. Comparison of Optimal DG Allocations for Different Optimization Models and Workflows

This subsection presents the optimal allocations of DGs obtained according to different optimization models and workflows used in the research study. The DG locations (bus) in the network and nominal DG apparent powers (S_n) are presented in Table 9 to give a comparable overview of the obtained solutions of the optimization problem, according to the different research scenarios used in the study.

Table 9. Results comparison for optimization models and workflows in case of single-objective optimizations.

Scenarios–Workflows Objective		Opt 1—WF1 Opt 2—WF2 Opt 3—WF2 W_{loss}	Opt 1—WF1 Opt 2—WF2 Opt 3—WF2 W_S	Opt 2 WF1 W_{loss}	Opt 2 WF1 W_S	Opt 3 WF1 W_{loss}	Opt 3 WF1 W_S
PV DG	bus	684	611	634	-	684	-
	S_n (kVA)	120	17	86	-	122	-
Wind DG	bus	634	611	634	611	632	632
	S_n (kVA)	258	120	228	55	936	745
Bio-gas DG	bus	692	670	692	670	692	671
	S_n (kVA)	1584	1810	1158	1637	1205	1242

4.5. Impact of Input Data Resolution on the Optimization Problem Solution

The previous procedures used in the presented research are analyzed to propose a suitable optimization model for the next steps in the study. Based on the results presented in the three previous subsections, the optimization model Opt 1 is chosen as the one with the best results (more details about the choice are given in the Discussion section below) for the application in the rest of the study. In this subsection, the results of the performing research step presented in Figure 6 are presented. The purpose of this step is to investigate the impact of the input data resolution on the solution of the optimization problem. The solutions for yearly input data with daily and monthly resolutions are compared against the solutions for the data with hourly resolution (as the base case). The input data with daily and monthly resolutions are generated as described in Section 4.1. Table 10 shows objective function values without installed DGs for different resolutions of input data. In Tables 11 and 12, an overview of solution comparisons presented in Figure 6 are shown for optimization model Opt 1 and single optimization problems (with W_{loss} and W_S objectives separately).

Table 10. The objective function values without installed DGs for the used input data resolutions.

Input Data Resolution	Hourly 8760 Data	Daily 365 Data	Monthly 12 Data
W_{loss} (kWh)	39,859	28,967	27,066
W_S (MVAh)	5555	5528	5516

Table 11. Impact of the input data resolution (Opt 1 optimization problem with W_{loss} objective).

Input Data Resolution	PV DG Allocation Bus-Size (kVA)	Wind DG Allocation Bus-Size (kVA)	Bio-Gas DG Allocation Bus-Size (kVA)	W_{loss} (kWh)	W_{loss} Decreasing in (%)
Hourly data 8760 data	684–120	634–258	692–1584	8400	78.93
Daily data 365 data	684–313	634–328	671–826	5136	82.27
Monthly data 12 data	646–419	634–239	692–614	4353	83.24

Table 12. Impact of the input data resolution (Opt 1 optimization problem with W_S objective).

Input Data Resolution	PV DG Allocation Bus-Size (kVA)	Wind DG Allocation Bus-Size (kVA)	Bio-Gas DG Allocation Bus-Size (kVA)	W_S (MVAh)	W_S Decreasing in (%)
Hourly data 8760 data	611–17	611–120	670–1810	288.27	94.81
Daily data 365 data	-	634–310	671–1060	164.11	97.03
Monthly data 12 data	-	634–150	671–691	140.10	97.46

4.6. Robustness of the Proposed Optimization Model

Based on the presented results, the optimization model Opt 1 is proposed as a framework for optimal DG allocation and power management. Let us recall that Opt 1 model results with solutions of the optimal allocation of DG, constant optimal power factors of DGs, and trained ANN for power management of the controllable DGs (Figure 1). After the optimal solution with hourly input data is found, the optimal allocation of uncontrollable DGs (PV and wind plants) and the trained ANN is implemented in the simulation model to drive the output power of the controllable DG (the bio-gas plant) with input data with higher resolution than the one used during the optimization process. The 15 min resolution data are used here, applying two scenarios. The ANN inputs are normalized PV and wind plant power outputs and load shape values. The 15 min PV and wind plant outputs, as well as the load shapes, are generated according to the procedure given in Section 4.1. Both scenarios assume the known real PV and wind plant power outputs. The one scenario (S1) assumes unknown real load shape values, in this case, the load shape ANN inputs are defined, according to the given load shapes (used during the optimization) and different from the real load shape value. In this scenario, for each four real load data in the load shape, the load inputs to ANN are the same (according to the given hourly load shape). The other scenario (S2) assumes known load shape values are used as ANN inputs. Both scenarios S1 and S2 are visualized in Figure 7. The results for these two scenarios for ±20% uncertainty (according to (3)) of load shapes and the DG profile are given in Table 13.

Table 13. Results of the proposed model application for 15 min input data for Opt 1 model—uncertainty of 20%.

Load Shape Values for ANN Input	W_{loss} no DG (kWh)	W_{loss} with DG (MVAh)	Losses Decreasing (%)
According to given load shapes	21,110	4748	77.51
Real values	21,110	4670	77.88
Difference	-	78	−0.07

4.7. Implementation on Middle Sized Distribution Network—IEEE 37 Node Test Feeder

In this subsection, the proposed procedure is applied on the IEEE 37 Node Test Feeder as an example of the distribution network of middle size. Based on the results and analysis of the procedure implementation on the IEEE 13 Node Test Feeder (given in previous subsections), the optimization model Opt 1, workflow WF 1, yearly profiles with hourly data and single objective (W_{loss}) optimization are applied on the IEEE 37 network. The obtained results are presented in Tables 14 and 15. The two optimization processes are performed: one for allocation of a total of three DGs (1xPV DG, 1xW DG, 1xBG DG), and the second for a total of six DGs (2xPV DG, 2xW DG, 2xBG DG).

Table 14. Optimal allocation of DGs in the IEEE 37 Node Test Feeder.

Input Data Resolution	PV DG Allocation bus-Size (kVA)	Wind DG Allocation bus-Size (kVA)	Bio-Gas DG Allocation bus-Size (kVA)
One of each DG 1xPV DG, 1xW DG, 1xBG DG	744–70	737–156	734–1000
two of each DG 2xPV DG, 2xW DG, 2xBG DG	712–78 704–86	737–141 701–74	728–49 701–203

Table 15. Decreasing the yearly power losses by optimal allocation and power management of DGs in the IEEE 37 Node Test Feeder.

Number of DGs	W_{loss} no DG (kWh)	W_{loss} with DG (MVAh)	Losses Decreasing (%)
3-1xPV DG, 1xW DG, 1xBG DG	10,755	4981	53.69
6- 2xPV DG, 2xW DG, 2xBG DG	10,755	4880	54.63

5. Discussion

Due to the extensive material (caused by many steps in the proposed research framework) presented in the previous section, comments about the obtained results are given in this section instead of in the previous one. The comments below are sorted, concerning the specific framework step referred to in Sections 4.2–4.6. The results obtained by investigation of the different optimization models and used workflow (Section 4.2) indicate that the simplest proposed model (Opt 1) gives the best results. The Opt 1 model finds optimal allocations of DGs, fixed values of DG power factors, and the trained ANN that generates the production profile of the controllable DG. Comparing results of single-objective and multiobjective optimizations (Table 4), it can be stated that the single-objective approach finds a better solution than two-objective optimization for the WS objective. In the case of single-objective optimization with the Wloss function, the solution is the same as that obtained by the two-objective optimization. Comparing the solutions obtained for different optimization models (Opt 1, Opt 2, and Opt 3) in Tables 4–6, unexpected results occur. As mentioned before, the best results are obtained for the Opt 1 model, but the authors expected models Opt 2 and Opt 3 to be better. In the Opt 2 and Opt 3 models, the DG power factors are also controlled (variable in time); the hypothesis was that this scenario would give better solutions. There is a need for further research to determine the reason for these unexpected results. Comparing the results in Tables 7 and 8 with those in Tables 4–6, it can be concluded that the initial solution has a low impact on the solution quality. The results presented in Tables 10–12 emphasize the importance of the input data resolution. Except for the difference in objective function absolute amounts, the important difference is in the optimal DG allocations obtained for different data resolution. The results in Table 13 show the possibility of the proposed framework application in the case of using real data with a resolution higher than that used in the optimization model tuning. The input of different data into ANN give close obtained objective values.

As mentioned before in the Introduction section, there are a few research studies that considered all three optimization problem issues (time changes of load and production profiles, co-simulation approach, together with optimization of the optimal allocation and power control of DGs) simultaneously. It is difficult to compare the research results with the existing studies directly because different objective functions and distribution power networks are used in the literature. Ref. [6,7] considered variable load and production profiles without a co-simulation approach, and only optimal allocation of DGs was solved. In Ref. [6], the example of the microgrid (from the literature) is used to present the results of the DG power outputs optimization (with DG locations given in advance). The objective functions are the minimization of operational costs and pollutants emission. The presented results show conflicted objective values in ranges of USD 760–870 and 960–1115 kg for the operational costs and the emission amount, respectively. In [7], the objective function is losses minimization, and the proposed method is applied to the example of the IEEE 14 bus test network. The obtained results show a decrease in the network losses to about 38% of losses for the basic case (without installed DGs). Refs. [9,11] deal with the co-simulation approach to find the optimal allocation of DGs. Ref. [9] solves the optimal allocation problem (without considering the DGs power management) using constant loads, and the proposed method is applied on the IEEE 37 node test network. The objectives are minimizing the nodal voltage variations and installation costs of DGs. The results show a reduction in value of the objective function to about 64% of its initial value (without DGs). In [11], the objective function is minimizing the network power losses, and the constant load and DG outputs are considered. The proposed procedure is applied to the IEEE 123 bus distribution network. The obtained results give about 79% power loss reduction of the initial losses (without DGs). As stated above in the Introduction section, the research study presented in [20] is closest to the research presented here. In [20], the variable load and DG production profiles, as well as optimal allocation and power management, are considered. The external simulation tool is used to calculate energy loss as the objective function. The presented method is applied to a power distribution network consisting of 69 buses. The presented results show power loss reductions in ranges (depending on numbers of DGs) of 63–69% and 89–98% of the initial power losses (with no DGs) for constant load and the unity power factor and constant load and the optimized power factor, respectively. In the scenario, with variable load, the energy loss reduction is in the range (depend on objectives impacts in the objective function) of 72–95% of the initial energy loss. Ref. [20] also considers the active energy infeed from the upstream network, and the obtained results for this objective are in the range of 60–90%, reducing the basic case value (without installed DGs). Because the study [20] considers similar problem aspects as those in this research (the initial data and tested network are not the same), the research presented here can be relatively compared with [20]. The results presented here (Sections 4.2–4.6) show the next obtained values. The range (depending on the applied optimization model Opt 1–Opt 3) of energy loss reduction is 60–79% of values with no installed DGs. The reduced exchanged apparent energy is in the range of 74–94% of the amount without DGs. The proposed method shows an energy loss reduction for 15 min resolution data of 77% of the initial value. This shows the applicability of the method in the case of using input data different from the data used in the optimization procedure.

Further research will be directed to investigation procedures for the estimation of DG power factor controls to additionally increase the optimal solution quality.

6. Conclusions

The presented framework for optimal allocation and power management of DGs emphasizes the importance of the resolution of the input data that needs to be considered during the optimization process. The proposed estimation of the controllable DG output by the ANN significantly decreases the number of the decision variables in the optimization problem, especially in the case of the high input data resolution. The results obtained for the case study indicate that knowing the hourly input data can be used to successfully tune the optimal model, which can be used later with increased input data resolution (15 min data).

This research study, compared to the existing literature, investigates the problem of the optimal allocation and power management of DG, makes contributions considering all three aspects of the problem detected in the Introduction section (load and DGs variable profile, co-simulation approach, and simultaneous consideration of the optimal allocation and power management of DGs). As stated in the Introduction section, there are a few research papers that consider these three problem aspects, simultaneously. Only (considering here the reviewed literature) in [20] did the authors apply variable profiles of load and DG production, external software for the calculation of the objective values, and variable DG power factor (optimized for optimal allocation determined in advance) to manage the DG output. Unlike the existing literature on the topic, the research presented here dealt with simultaneous optimization of the DG allocation and power management, considering the yearly (with hourly resolution) load and DG production profiles, using a co-simulation approach. Besides this, the research proposed the application of ANN to manage DG outputs, which significantly decreases the number of decision variables that appear when yearly profiles are used.

The presented solution framework shows that it is possible to optimize the allocation and variable power outputs of DGs simultaneously in the case of high resolution input data. The high resolution of input data over a long time span (a year) produces a very high number of decision variables that need to be optimized. The demonstrated application of the ANN makes it possible to significantly decrease the number of decision variables with simultaneous consideration of the optimal allocation and power management of the DGs. For successful optimization of the DG power factor management, additional investigations of the procedures are required, which will be included in the optimization process simultaneously with the here-applied problem aspects.

Author Contributions: Conceptualization, M.B. and T.V.; methodology, T.V., M.B. and T.B.; software, V.J.Š., M.B. and T.V.; validation, T.V. and V.J.Š.; formal analysis, T.B. and V.J.Š.; writing—original draft preparation, M.B.; writing—review and editing V.J.Š., T.V., T.B. and M.B.; project administration, M.B. and funding acquisition, M.B. All authors have read and agreed to the published version of the manuscript.

Funding: This work was supported in part by the Croatian Science Foundation under the project number UIP-05-2017-8572.

Data Availability Statement: The data used and obtained in this study are available on https://drive.google.com/drive/folders/1QUFNBe1DykyVcmEboLFENeoW3bhfPGrx?usp=sharing (accessed on 12 March 2021).

Conflicts of Interest: The authors declare no conflict of interest. The founders had no role in the design of the study; in the collection, analyses, or interpretation of data; in the writing of the manuscript, or in the decision to publish the results.

Abbreviations

The following abbreviations, quantity and variables are used in this manuscript:

ABC	Artificial Bees Colony
ALO	Ant Lion Optimizer
BESS	Battery Energy Storage System
BG	Bio-gas
DE	Differential Evolution
DG	Distributed Generation
EV	Electrical Vehicle
GA	Genetic Algorithm
GWO	Grey Wolf Optimizer
HAS	Harmony Search Algorithm
IEEE	Institute of Electrical and Electronics Engineers
IN 1	Input data, hourly resolution
IN 2	Input data, daily resolution
IN 3	Input data, monthly resolution
LP 01	Type 1 of load profile/shape
LP 02	Type 2 of load profile/shape
LP 03	Type 3 of load profile/shape
MICP	Mixed-Integer Conic Programming
Opt 1	Optimization model/scenario 1
Opt 2	Optimization model/scenario 2
Opt 3	Optimization model/scenario 3
PSO	Particle Swarm Optimization
PV	Photovoltaic
SA	Sensitivity Analysis
SFL	Shuffled Frog Leap
THD	Total Harmonic Distortion
VPP	Virtual Power Plant
W	Wind
WF1	Workflow 1
WF2	Workflow 2
F	Multiobjective function
f_1	Objective 1 in the multiobjective function F
f_2	Objective 2 in the multiobjective function F
H_1	The first neuron in the hidden layer
H_{n+k+m}	The last neuron in the hidden layer
k	Number of PV plants
$I_{k,e}$	Calculated current in k-th network line
$I_{k,max}$	Rated/allowed current of k-th network line
j	Number of BG plants
z	relative number used in generating the random variable
LF_{15min}	Load factor for calculating 15 min load profile/shapes
LF_h	Hourly load factor value
$LS_{1,i}$	i-th data in the first load shape
$LS_{n,i}$	i-th data in n-th load shape
m	Number of wind plants
n	Number of load shapes
N	Number of data in load shapes (same as number of time steps) and DG production profiles
$P_{BG,1,i}$	i-th output of the first BG plant
$P_{BG,j,i}$	i-th output of the j-th BG plant
$P_{exc,i}$	Active power exchange at the i-th time step
$P_{loss,i}$	Total network active power losses at the i-th time step
$P_{PV,1,i}$	i-th data in the first production profile of PV plant
$P_{PV,k,i}$	i-th data in k-th production profile of PV plant

$P_{wind,1,i}$	i-th data in the first production profile of wind plant
$P_{wind,m,i}$	i-th data in m-th production profile of wind plant
$Q_{exc,i}$	Reactive power exchange at the i-th time step
t_i	Duration of the i-th time step
$\mathcal{U}$	Probability density function of the uniform distribution
$V_{i,e}$	Calculated nodal voltage in i-th network bus
V_{max}	Maximum of the nodal voltage value
V_{min}	Minimum of the nodal voltage value
W_{loss}	Energy of yearly losses as one of objectives in F
W_S	Apparent yearly energy exchanged between the distribution and upstream network
x	Decision variable vector
X	The random number generated according to $\mathcal{U}$
x_{lb}	Lower bounds of the decision variables
x_{ub}	Upper bounds of the decision variables

References

1. Cui, H.; Dai, W. *Multi-Objective Optimal Allocation of Distributed Generation in Smart Grid*; IEEE: Yichang, China, 2011; pp. 713–717. [CrossRef]
2. Nasiraghdam, H.; Jadid, S. Load model effect assessment on optimal distributed generation (DG) sizing and allocation using improved harmony search algorithm. In Proceedings of the 2013 Smart Grid Conference (SGC), Tehran, Iran, 17–18 December 2013. [CrossRef]
3. Moeini, A.; Kamwa, I.; de Montigny, M. *Optimal Multi-Objective Allocation and Scheduling of Multiple Battery Energy Storages for Reducing Daily Marginal Losses*; IEEE: Washington, DC, USA, 2015; pp. 1–5. [CrossRef]
4. Kumar, D.S.; Tianyi, H.; Srinivasan, D.; Reindl, T.; Shenoy, U.J. Optimal distributed generation allocation using evolutionary algorithms in meshed network. In Proceedings of the 2015 IEEE Innovative Smart Grid Technologies—Asia (ISGT ASIA), Bangkok, Thailand, 4–6 November 2015; IEEE: New York, NY, USA, 2015. [CrossRef]
5. Ju, L.; Li, H.; Zhao, J.; Chen, K.; Tan, Q.; Tan, Z. Multi-objective stochastic scheduling optimization model for connecting a virtual power plant to wind-photovoltaic-electric vehicles considering uncertainties and demand response. *Energy Convers. Manag.* **2016**, *128*, 160–177. [CrossRef]
6. Hosseini, K.; Araghi, S.; Ahmadian, M.B.; Asadian, V. Multi-objective optimal scheduling of a micro-grid consisted of renewable energies using multi-objective Ant Lion Optimizer. In Proceedings of the 2017 Smart Grid Conference (SGC), Tehran, Iran, 20–21 December 2017. [CrossRef]
7. Xuemei, S.; Bin, Y.; Xuyang, W.; Jin, Y.; Ciwei, G. Study on Optimal Allocation of Distributed Generation in Urban and Rural Distribution Network Considering Demand Side Management. In Proceedings of the 2017 International Conference on Smart Grid and Electrical Automation (ICSGEA), Changsha, China, 27–28 May 2017; pp. 560–566. [CrossRef]
8. Sanjay, R.; Jayabarathi, T.; Raghunathan, T.; Ramesh, V.; Mithulananthan, N. Optimal Allocation of Distributed Generation Using Hybrid Grey Wolf Optimizer. *IEEE Access* **2017**, *5*, 14807–14818. [CrossRef]
9. Kim, I. Optimal distributed generation allocation for reactive power control. *IET Gener. Transm. Distrib.* **2017**, *11*, 1549–1556. [CrossRef]
10. Phawanaphinyo, P.; Keeratipranon, N.; Khemapatapan, C. Optimal Active Power Loss with Feeder Routing Collaborate Distributed Generation Allocation and Sizing in Smart Grid Distribution. In Proceedings of the 2017 International Conference on Economics, Finance and Statistics (ICEFS 2017), Hong Kong, 14–15 January 2017; pp. 387–392. [CrossRef]
11. Kumawat, M.; Gupta, N.; Jain, N.; Bansal, R. Optimally Allocation of Distributed Generators in Three-Phase Unbalanced Distribution Network. *Energy Procedia* **2017**, *142*, 749–754. [CrossRef]
12. da Rosa, W.M.; Teixeira, J.C.; Belati, E.A. New method for optimal allocation of distribution generation aimed at active losses reduction. *Renew. Energy* **2018**, *123*, 334–341. [CrossRef]
13. LIU, H.; XU, L.; ZHANG, C.; SUN, X.; CHEN, J. *Optimal Allocation of Distributed Generation Based on Multi-Objective Ant Lion Algorithm*; IEEE: Chengdu, China, 2019; pp. 1455–1460. [CrossRef]
14. Home-Ortiz, J.M.; Pourakbari-Kasmaei, M.; Lehtonen, M.; Mantovani, J.R.S. Optimal location-allocation of storage devices and renewable-based DG in distribution systems. *Electr. Power Syst. Res.* **2019**, *172*, 11–21. [CrossRef]
15. Karunarathne, E.; Pasupuleti, J.; Ekanayake, J.; Almeida, D. Optimal Placement and Sizing of DGs in Distribution Networks Using MLPSO Algorithm. *Energies* **2020**, *13*, 185. [CrossRef]
16. Hassan, A.S.; Sun, Y.; Wang, Z. Multi-objective for optimal placement and sizing DG units in reducing loss of power and enhancing voltage profile using BPSO-SLFA. *Energy Rep.* **2020**, *6*, 1581–1589. [CrossRef]
17. Gampa, S.R.; Jasthi, K.; Goli, P.; Das, D.; Bansal, R. Grasshopper optimization algorithm based two stage fuzzy multiobjective approach for optimum sizing and placement of distributed generations, shunt capacitors and electric vehicle charging stations. *J. Energy Storage* **2020**, *27*, 101117. [CrossRef]
18. HA, M.P.; Nazari-Heris, M.; Mohammadi-Ivatloo, B.; Seyedi, H. A hybrid genetic particle swarm optimization for distributed generation allocation in power distribution networks. *Energy* **2020**, *209*, 118218. [CrossRef]

19. Lim, K.Z.; Lim, K.H.; Wee, X.B.; Li, Y.; Wang, X. Optimal allocation of energy storage and solar photovoltaic systems with residential demand scheduling. *Appl. Energy* **2020**, *269*, 115116. [CrossRef]
20. Huy, P.D.; Ramachandaramurthy, V.K.; Yong, J.Y.; Tan, K.M.; Ekanayake, J.B. Optimal placement, sizing and power factor of distributed generation: A comprehensive study spanning from the planning stage to the operation stage. *Energy* **2020**, *195*, 117011. [CrossRef]
21. Emmerich, M.T.M.; Deutz, A.H. A tutorial on multiobjective optimization: Fundamentals and evolutionary methods. *Nat. Comput.* **2018**, *17*, 585–609. [CrossRef] [PubMed]
22. Schlueter, M.; Erb, S.O.; Gerdts, M.; Kemble, S.; Rückmann, J.J. MIDACO on MINLP space applications. *Adv. Space Res.* **2013**, *51*, 1116–1131. [CrossRef]
23. Schlüter, M.; Egea, J.A.; Banga, J.R. Extended ant colony optimization for non-convex mixed integer nonlinear programming. *Comput. Oper. Res.* **2009**, *36*, 2217–2229. [CrossRef]
24. Abadi, M.; Agarwal, A.; Barham, P.; Brevdo, E.; Chen, Z.; Citro, C.; Corrado, G.S.; Davis, A.; Dean, J.; Devin, M.; et al. TensorFlow: Large-Scale Machine Learning on Heterogeneous Systems. 2015. Available online: tensorflow.org (accessed on 12 March 2021).
25. Chollet, F. Keras. 2015. Available online: https://github.com/fchollet/keras (accessed on 12 March 2021).
26. Dugan, R.C.; McDermott, T.E. An open source platform for collaborating on smart grid research. In Proceedings of the 2011 IEEE Power and Energy Society General Meeting, Detroit, MI, USA, 24–28 July 2011. [CrossRef]
27. Distribution Test Feeder Working Group—IEEE PES Distribution System Analysis Subcommittee. Distribution Test Feeders. Available online: https://site.ieee.org/pes-testfeeders/resources/ (accessed on 15 May 2018).
28. Pflugradt, N.; Muntwyler, U. Synthesizing residential load profiles using behavior simulation. *Energy Procedia* **2017**, *122*, 655–660. [CrossRef]
29. Pfenninger, S.; Staffell, I. Long-term patterns of European PV output using 30 years of validated hourly reanalysis and satellite data. *Energy* **2016**, *114*, 1251–1265. [CrossRef]
30. Staffell, I.; Pfenninger, S. Using bias-corrected reanalysis to simulate current and future wind power output. *Energy* **2016**, *114*, 1224–1239. [CrossRef]

MDPI

Article

An Adaptive Protection Scheme for Coordination of Distance and Directional Overcurrent Relays in Distribution Systems Based on a Modified School-Based Optimizer

Mohamed Abdelhamid [1], Salah Kamel [1], Ahmed Korashy [1], Marcos Tostado-Véliz [2,*], Fahd A Banakhr [3] and Mohamed I. Mosaad [3]

1 Department of Electrical Engineering, Faculty of Energy Engineering, Aswan University, Aswan 81528, Egypt; mohammed.abdulhameed@aswu.edu.eg (M.A.); skamel@aswu.edu.eg (S.K.); ahmed.korashy2010@yahoo.com (A.K.)
2 Electrical Engineering Department, University of Jaen, EPS, 23700 Linares, Spain
3 Electrical and Electronic Engineering Technology Department, Yanbu Industrial College, Yanbu Al Sinaiyah, Yanbu 46452, Saudi Arabia; banakherf@rcyci.edu.sa (F.A.B.); habibm@rcyci.edu.sa (M.I.M.)
* Correspondence: mtostado@ujaen.es

Citation: Abdelhamid, M.; Kamel, S.; Korashy, A.; Tostado-Véliz, M.; Banakhr, F.A.; Mosaad, M.I. An Adaptive Protection Scheme for Coordination of Distance and Directional Overcurrent Relays in Distribution Systems Based on a Modified School-Based Optimizer. *Electronics* **2021**, *10*, 2628. https://doi.org/10.3390/electronics10212628

Academic Editors: Marinko Barukčić, Nebojša Raičević and Vasilija Šarac

Received: 9 September 2021
Accepted: 19 October 2021
Published: 27 October 2021

Abstract: This paper presents an adaptive protection scheme (APS) for solving the coordination problem that deals with coordination directional overcurrent relays (DOCRs) and distance relays second zone time, in relation to coordination with DOCRs. The coordination problem becomes more complex with the impact of renewable energy sources (RES) when added to the distribution grid. This leads to a change in the grid topology, caused by the on/off states of the distribution generators (DG). The frequency of topological changes in distribution grids poses a challenge to the power system's protection components. The change in the state of DGs leads to malfunction in reliability and miscoordination between protection relays, since that causes a direct effect to the short circuit currents. This paper used the school-based optimization (SBO) algorithm, which simulates the educational process, in order to deal with coordination problems. That algorithm is modified (MSBO) by modified both learning and teaching processes. The IEEE 8-bus test system and IEEE 14-bus distribution network are used to validate the proposed coordination system's effectiveness when dealing with the coordination process between distance and DOCRs, at both the near- and far-end in the typical topological grid and with DGs in working order.

Keywords: power system protection; overcurrent relays; protection relays; metaheuristic; school-based optimizer

1. Introduction

Nowadays, the protection field is one of the more indexing issues in power systems. Directional overcurrent relays (DOCRs) and distance relays are both commonly used for protecting transmission lines. These protection devices monitor the transmission lines from both ends of the lines for faults that cause trip scenarios to be activated.

Overcurrent relays (OCRs) generally work by the magnitude of the fault current, which is set inside relay's parameters, while in DOCRs, adding the direction of the passing current through transmission lines. This direction is determined by voltage phasor from the potential transformer. So, DOCRs are more expensive than normal OCRs but more effective than OCRs. These relays must be operating in the backup case, with a delay time higher than the primary relay [1].

The second protection is distance relays, which have two main zones. The first one works immediately after fault detection. This zone covers 80% of the transmission line to ignore calculation errors. Then, the second zone covers up to 120% of the transmission line by delay time; this wide area covers a part of another transmission line [2].

The main problem in this paper regards reducing the protection relay's operation times, to provide the ability for protection devices to isolate the fault area. This saves the lifetime of power system components, and the power system becomes healthier and more reliable. However, the coordination problem of DOCRs and distance relays is more complex and highly constrained, owing to constraints between DOCRs pairs and DOCRs and distance relays pairs. The miscoordination of these protection relays overlap protection operates and does not utilize the advantages of both distance and DOCRs relays [3,4].

The impact of RES-based DGs adds to the distribution system. RES, such as solar energy and wind energy sources, are integrated with the power system. DGs and the coordination problem present many challenges, such as the change of fault current magnitude and flow of direction [5].

This coordination challenge, which is the result of DGs, needs a flexible structure. This paper discussed adaptive protection systems (APS), in order to solve this protection coordination problem. APS gives the ability to change relays settings for both DOCRs and distance relays, according to change in network states, based on the DG's on/off states, using predetermined settings. APS was tested with various scenarios, which are probably tripped in-network, and the optimal settings for protection relays in each scenario were determined. This gives the protection system the ability to minimize miscoordination and malfunction. The main advantage of APS is making the protection system more selective and reliable than conventional or fixed systems [6]. APS's settings group of protection relays is determined by computing optimal settings using an optimization algorithm for each scenario, which is based on the DG's states [7].

In recent years, many optimization algorithms are used for solving coordination problems in literature, of DOCRs coordination, such as the particle swarm optimizer (PSO) and modified PSO in [8], genetic algorithm (GA) and hybrid GA in [9], biogeography-based optimization algorithms (BBO) in [10], differential evolution algorithm (DE) and trigonometric DE algorithm (Tri-DE) in [11], firefly algorithm (FA) and improved FA (IFA) in [12], hybridized whale optimization algorithm (WOA), and hybridized WOA in [13], Jaya Algorithm and oppositional Jaya algorithm (OJaya) in [14], moth–flame optimization (MFO) and improved MFO (IMFO) in [15], political optimization algorithm (PO) in [1], artificial optimizing algorithm(AEO) in [16], and evaporation rate water cycle algorithm in [17].

Then, for the coordination of both the DOCRs and distance, such as the genetic algorithm (GA) in [18,19], water cycle algorithm (WCA) [19], Jaya optimization algorithm [20], grey wolf optimization (GWO) [19], ant colony optimization (ACO), and hybrid ACO algorithm in [21].

The adaptive protection scheme is important for coordinating the protection relays, in order to deal with the change in topology of the distribution network, which results from the DG's on/off status. This topological change causes a change in the short-circuit current. Hence, modern protection systems, which deal with DGs or RES, are needed for an adaptive scheme.

APS is basically dependent on the communication network between the smart grid's components, as a part of information and communication technologies (ICT), or it is dependent on SCADA. These communication networks give APS the ability to set relays remotely.

Because of the real-time performance of the revolution of optimization algorithms (in terms of millisecond or microseconds), as well as high computerized performance, in many research papers, APS is shown to be dependent on the optimization algorithms to coordinate DOCRs, such as using the: particle swarm optimization (PSO) in [22], genetic algorithm (GA) in [23], differential evolution algorithm (DEA) in [24], ant colony optimization (ACO) [25], gravitational search algorithm (GSA) in [26], firefly algorithm (FA) in [27], manta ray foraging optimization (MRFO) in [7], and hybrid Harris hawks optimization (HHO) in [28].

Metaheuristic optimization algorithms usually generate random initial values, as its population within search space limiters then improves the population fitness within a systematic process. The standard of metaheuristic optimization algorithms is always formed by intrapopulation collaboration. The original SBO algorithm utilized subgroups of the parallel populations, with independent values that collaborate. Increase the capability of exploration of the algorithm and improve the overall efficiency. SBO is a collaborative, multi-population framework utilized by TLBO. This algorithm used two stages: the first stage is about a series of metaheuristics works, independent for exploring the different areas of the search space. Then, the second stage concentrated the search on the sub-region within the best solutions. This type of algorithm has many challenges; one of them is selecting and implementing the first stage termination criterion. The terminal criterion introduces parameters that need to be tuned for a specific problem [29].

SBO extends the basic model of TLBO, with both learning and teaching phases; however, MSBO used TLBO with a modified learning phase. Then, teachers can be rearranged with a roulette wheel role to other classrooms to share their knowledge; while, MSBO used multiple teachers for each classroom to improve share knowledge processes between classrooms and increased the exploitation of the population into the teaching phase [30].

SBO is applied to solve many other engineering optimization problems, such as steel frame design in [29,31] and solar cell parameters estimation in [32]. SBO is effective in solving these optimization problems.

Other methods are suggested to deal with APS, such as multi-agents in [33,34] and Q-learning with an environment APS in [35].

Contributions of this paper are as follows:

- An adaptive protection scheme was designed to coordinate both DOCRs and distance relays. This paper is the first one that deals with this problem in APS, as a solution to the DG impact. The effect of distance relays complicates this coordination problem in the DOCR's coordination process, in addition to the impact of DGs.
- The original SBO algorithm was modified to improve the response and convergence of the proposed algorithm. In doing so, two main points were modified, learning and the teacher-selected process. This algorithm could be effective in solving other power systems' important topics, such as load frequency control, parameter estimation of the solar cell, optimal location, and the sizing of DGs.
- The proposed protection system in this paper was tested on both IEEE 8-bus and IEEE 14-bus distribution networks, with the effect of DG's on/off states.

The rest of the paper is as follows: Section 2 is about the mathematical modelling of coordination problems. Section 3 presents the proposed protection scheme. Then, in Section 4, the performance of both SBO and MSBO, for solving the coordination problem in IEEE 8-bus and IEEE 14-bus distribution networks, is presented. Finally, Section 5 shows the conclusions.

2. The Mathematical Modelling of Coordination Problem

The main goal of this paper is to get the optimal coordination of DOCRs and distance relays. The optimal solution to this problem is minimizing the total operation time of DOCRs at both ends of the near-end (*TNR*) and far-end (*TFR*), in addition to the second zone time of distance relays (T_{Z2}). The minimum total operation time is the objective function (*OF*), shown as following [21,36,37]:

$$OF = min\left(\sum_{i=1}^{n} TNR_i + \sum_{i=1}^{n} TFR_i + \sum_{i=1}^{n} T_{Z2}i + F^{Pen}\right), \tag{1}$$

The standard time inverse DOCRs characteristics, depending on the international electrotechnical commission (IEC) standards, are presented by the following equation [16]:

$$T_i = \frac{\propto *TDS_i}{\left(\frac{I_f}{I_{pi}}\right)^{\beta} - \gamma}, \tag{2}$$

where T_i is the operation time of relay at any end of transmission line for i relay, TDS is its time dial setting, and I_p is its pick-up current. The other α, β, and γ are constants with 0.14, 0.02, and 1, respectively [1].

2.1. The Problem's Limiters

The main limiter of any protection relay is the maximum operation time (T_{max}) to prevent bad operation, which saves the power system component's lifetime. That limiter must be lower than 2 s [16].

Any relay settings, in coordination with the problem, have minimum and maximum limiters, as shown in the following equations [21]:

$$TDS_{min} \leq TDS \leq TDS_{max}, \tag{3}$$

$$Ip_{min} \leq Ip \leq Ip_{max}, \tag{4}$$

$$Tz2_{min} \leq Tz2 \leq Tz2_{max}, \tag{5}$$

2.2. The Problem's Constraints

The proposed optimization problem becomes a higher constraint problem, via the constraints between the primary and backup pair of DOCRs, in addition to the relationship between the DOCRs, distance, and pairs relay at both ends (near and far). Those constraints are used to avoid miscoordination, which may happen during faults between protection relays.

The relationship between DOCRs pair relays, at any end, as shown in Figure 1, must deal with the backup relay (t_b), operated with a delay on the primary relay (t_p). This delay time is called coordination time interval (CTI). The value of CTI is determined according to the type of protection relays. For electromagnetic relays, the CTI value must be more than 0.3 s, while, in the case of digital relays, more than 0.2 s; the digital relays are used in this paper [38]. The following equation shows these constraints [21]:

$${t_b}^{f1} - {t_p}^{f1} > CTI, \tag{6}$$

$${t_b}^{f2} - {t_p}^{f2} > CTI, \tag{7}$$

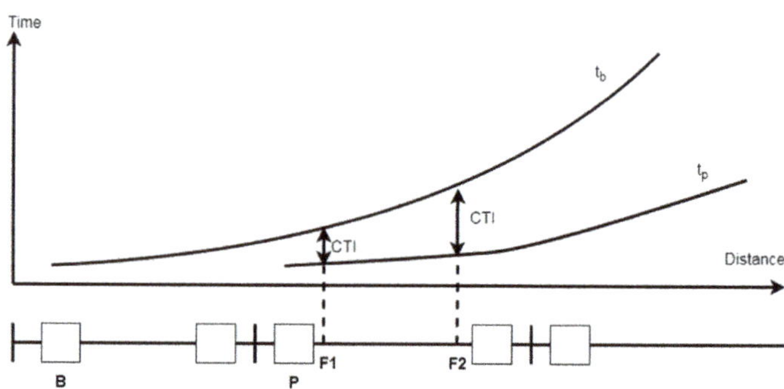

Figure 1. The relationship between primary and backup DOCRs.

The relationship between DOCRs and distance pair relays is shown in Figure 2. The backup distance relay aliasing, with the primary DOCRs relay at the near end, and T_{Z2b} must delay on ${t_p}^{f1}$, with the CTI as described in Equation (8); Equation (9) describes the relationship between distance and DOCRs at the far end. At the far end, the second zone of primary distance relay (T_{Z2p}) must delay on primary DOCRs operation time $\left({t_p}^{f1}\right)$ with CTI [21].

$$T_{Z2b} - {t_p}^{f1} > CTI, \tag{8}$$

$$T_{Z2p} - t_p{}^{f2} > CTI, \tag{9}$$

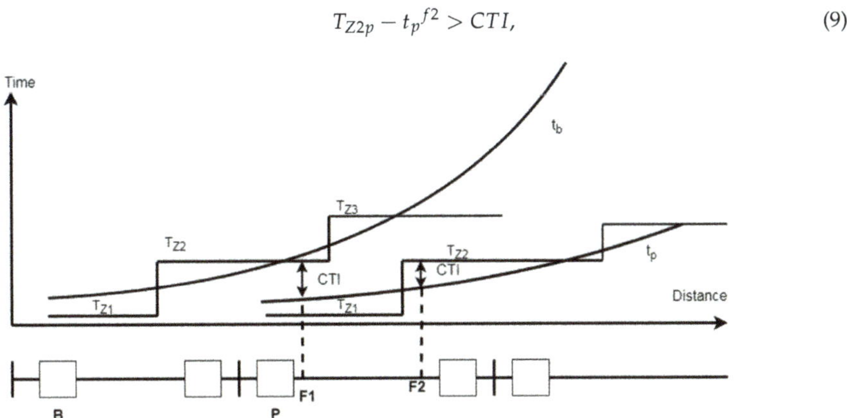

Figure 2. The relationship between DOCRs and Distance pair relays.

This relationship was developed to specify the minimum value of the second zone of the distance relay, based on the operation time of the primary relay at both ends near and far. This idea is discussed in [39]. Equations (8) and (9) are rearranged to Equations (10) and (11). Then, the maximum value of these equations is used as the specific second zone of distance relay's time. This point helps to reduce the penalty and constraints.

$$T_{Z2b} = t_p{}^{f1} + CTI, \tag{10}$$

$$T_{Z2p} = t_p{}^{f2} + CTI, \tag{11}$$

$$T_{Z2} = max(T_{Z2b}, T_{Z2p}), \tag{12}$$

The penalty function is recommended for use in the main goal of eliminating miscoordinations, as in the following equation [40]:

$$F^{pen} = \mu * \begin{cases} 1\, if\ T^{backup} - T^{primary} < CTI \\ 0\, if\ T^{backup} - T^{primary} \geq CTI \end{cases}, \tag{13}$$

When miscoordination occurs in this penalty function, F^{pen} increases the total time of OF. As a result, the optimization algorithm attempts to eliminate miscoordination, in order to reduce the size of OF; μ is the weighting factor in this penalty function [37].

3. The Proposed Protection Scheme

3.1. Smart Grid and Adaptive Protection Scheme (APS)

In this research work, the proposed scheme is based on optimization solutions by an optimization technique. In this paper, the school-based optimization algorithm, used to evaluate the optimization solutions, in addition to this paper, included modifications for that algorithm, in order to improve its convergence characteristics and ability to find better optimization solution, as described in the next section.

The flow diagram (Figure 3) presents APS, considering DG's impact. The centralized processing server is used to optimize SCADA data. These data will be generated by APS-proposed algorithms, for resetting the DOCRs and distance relays. The following steps refer to the main points of the proposed APS flow chart.

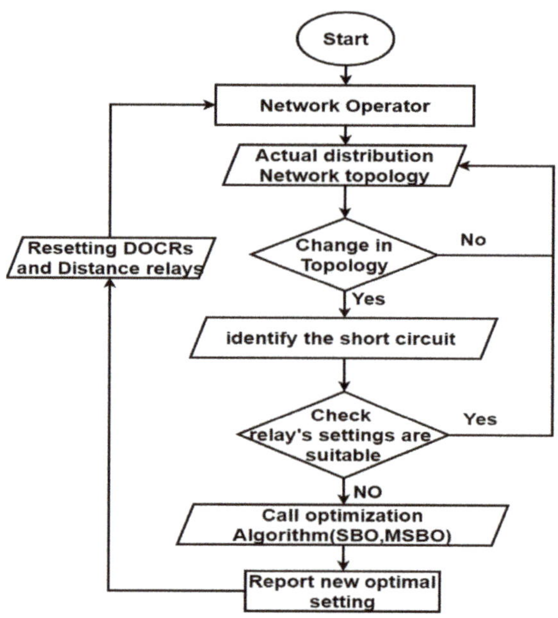

Figure 3. Flow diagram of APS.

The first point determined the actual distribution network topology, especially the state, location, and size of the DGs. Check for a change of distribution network topology. In case of no change, the APS stays with the current protection relay's settings; in the case of a change in topology, the APS moves to the next point.

In the second point, APS identifies the short current through CBs, ETAP was used in this paper for this mission. Then, check the ability of the current relay's settings, in order to save the protection system without loss-coordination of protection relays or miscoordination between protection relays. In case of the ability of the current setting to protect the distribution network, the APS returns to the previous point. However, it will move to the next point in the case that the relay's setting misses their job to protect the distribution network.

In the third point, APS calls the proposed optimized algorithm. Then, the algorithm searches for optimal solutions that are suitable to cover the changes in the distribution network, without miscoordination or loss-coordination. Finally, the APS reports the optimal solution of protection relay's settings and sends them through ICT and in-distribution network update IEDs. APS will stick to new changes in the distribution network [41].

3.2. Original School-Based Optimization Algorithm

SBO is a metaheuristic algorithm, as shown in its flowchart in Figure 4. SBO is formed from many classrooms and has many teachers. Each classroom used the TLBO algorithm, in order to be built. Each classroom has a teacher, which is the population with the best fitness. Teachers are joining to a pool of teachers. In this pool, teachers are distributed by a roulette wheel to a new classroom, in order to transfer the knowledge between them.

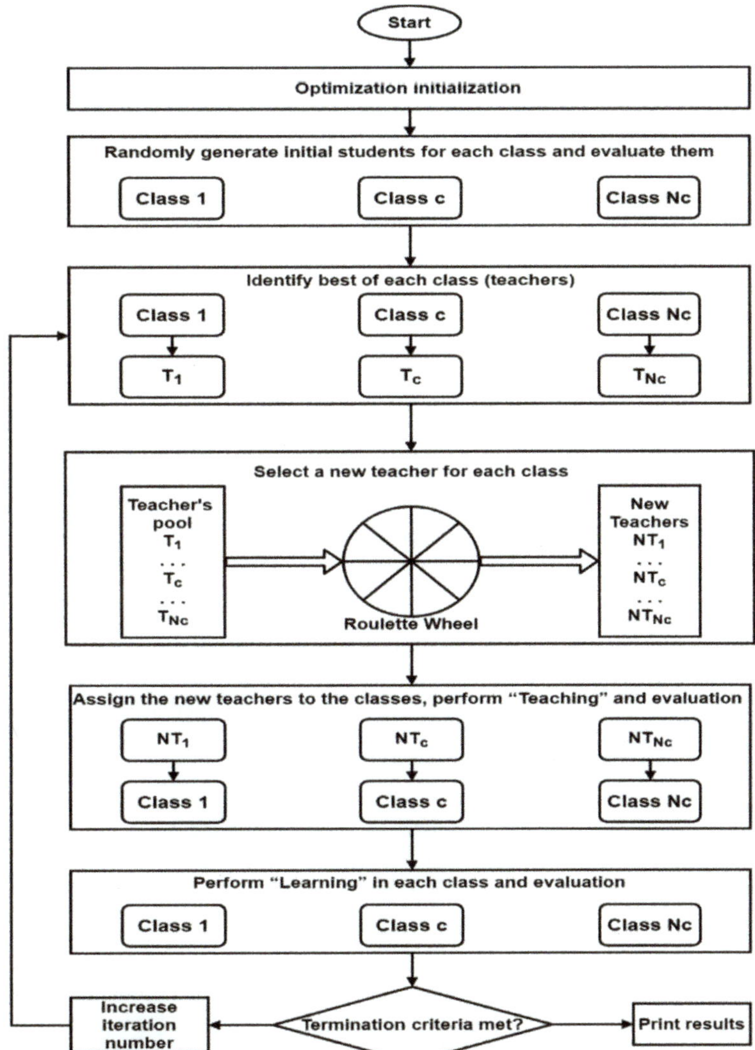

Figure 4. Flow chart of SBO algorithm.

The TLBO algorithm was inspired by the educational process in the classroom. That algorithm has two main phases. These phases are about the educational process and exchanging knowledge. The first phase is called the teacher phase. In this phase, the knowledge is transferred from teacher to students. The other phase is called the learning phase. That phase simulates the cooperative learning between students [29].

3.2.1. Teaching Phase

In this phase, the optimization algorithm simulates the teaching process to students who are trying to update themselves by knowledge transfer from their teacher. That representation is mathematically as follows:

$$X^k_{new}(j) = X^k_{old}(j) \pm \Delta(j)\,, \tag{14}$$

$$\Delta(j) = T_F \times r|M(j) - T(j)|, \tag{15}$$

where $X^k(j)$ refers to the solution as a student with index jth, $\Delta(j)$ is the difference between the teacher and the mean of the class, T_F is a teaching factor and equal to 2, r is a random value between [0, 1], $T(j)$ is the solution as a teacher, and $M(j)$ is the mean of the classroom and is represented as follows [30]:

$$M(j) = \frac{1}{N}\sum\nolimits_{k=1}^{N} X^k(j), \tag{16}$$

$$M(j) = \frac{\sum_{k=1}^{N} \frac{X^k(j)}{F^k}}{\sum_{k=1}^{N} \frac{1}{F^k}}, \tag{17}$$

where N is the population. F^k is the penalized fitness of student solution with indexing kth.

Equation (17) is about the fitness-based mean. This formula gives more emphasis to students and improves the performance of the TLBO algorithm [42].

At the end of the iteration, the solution that has the best fitness is chosen as a new teacher in the next iteration.

3.2.2. Learning Phase

Interactive learning between students in each classroom can develop the student's performance then develop the performance of the classroom. The learning phase is given by following steps:

1. Randomly selected student p and another q while $p \neq q$.
2. If the fitness of student p is better than student q.

$$X^p_{new}(j) = X^p_{old}(j) + r\left[X^p_{old}(j) - X^q(j)\right], \tag{18}$$

Otherwise,

$$X^p_{new}(j) = X^p_{old}(j) + r\left[X^q(j) - X^p_{old}(j)\right], \tag{19}$$

where r is a random number between [0, 1].

This phase moves student p towards student q if student q has a better solution; while, if student p has a better solution, it will move away from student q [42].

3.3. *Modified SBO Algorithm*

The original SBO algorithm was modified, as shown in its flowchart in Figure 5. That is based on two main points, in order to improve its ability to explore and exploit in the original algorithm. The first point is about the learning phase, which was discussed in the previous section. This part is modified by changing the techniques of learning between students by three additional steps, discussed below. The second point is to select more than one teacher, via a roulette wheel, in order to speed the process up and obtain a better knowledge transfer process inside the class (and with other classes).

3.3.1. First Point: A Modified Learning Phase

This point has been modified to exploit students in each classroom to reach new points within a limited search area that has better fitness values, by using the following steps:

1. This step does not choose the student p randomly but as the place in the classroom.
2. The student q is chosen randomly but not repeated with another student and cannot be equal to student p.
3. Equations (18) and (19) are both applied and choosing the best value to compare with the student p fitness value (to replace or not).

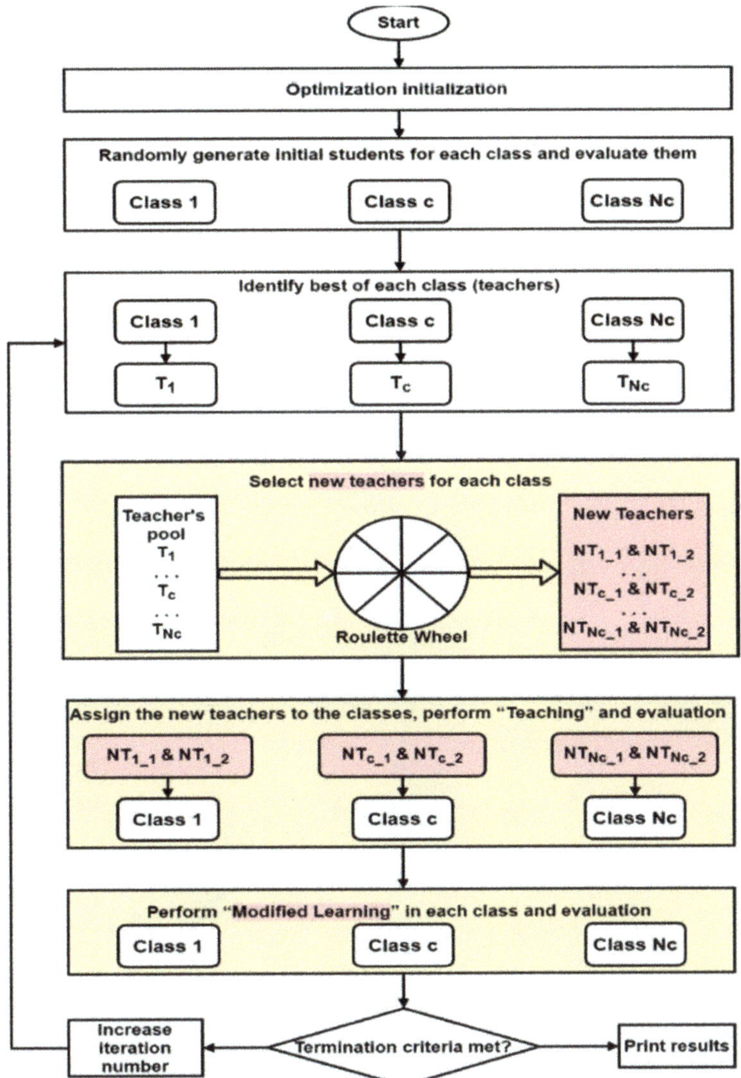

Figure 5. Flow chart of MSBO algorithm.

3.3.2. Second Point: Teacher Selected

The original algorithm selected one teacher, but the improved algorithm selected more than one. This point increased the knowledge transfer process between classrooms. In the original algorithm selected, one teacher from another classrooms carried knowledge from it; however, after that, each classroom takes knowledge from many classrooms and then selected new points after evaluating the fitness values, in order to accept the best fitness value between new points affected by teachers. This improved the teacher phase's equation used previously but repeated with each teacher. Teachers were distributed to classrooms by roulette wheel and this modified was used, too, but selected many teachers for each classroom. The number of teachers for each classroom was selected by users and, in this paper, double teachers were selected.

4. Results and Discussion

The optimization setting of TDS, IP, and TZ2 were tuned by the MATLAB program, used for both SBO and MSBO to solve the optimization problem. These algorithms, discussed in previous sections, used population, classrooms, and maximum iterations, with values are 300, 25, and 1000, respectively. The algorithms were successfully tested in coordination tested systems, i.e., the IEEE 8-bus test system and the IEEE 14-bus distribution network. Each test system has two varying cases: the first is the normal topological grid and the second is after added external power generation for the original grid.

The optimum settings were used to calculate the operation time of the primary and backup protection relays at the near-end and the far-end. These points were tested to succeed in the optimal solution, in order to pass system constraints. The protection devices, assumed with digital relays and CTI, must be greater than or equal to 0.2 s. Relays with normal characteristics constants are α, β, and γ, with values of 0.14, 0.02, and 1.0, respectively, maximum and minimum TDS values of 1.1 s and 0.1 s, respectively, maximum and minimum PS were 4 and 0.5, respectively, and the maximum time for operating the primary DOCRs or distance relays was 1.5 s [21].

MATLAB R2016a was used on a computer with a CPU of 1.70 GHz processor and 4 GB DDR3 RAM, for tuned optimum settings, while ETAP 12.6.0 was used for the calculation three phase fault currents.

4.1. Test System I: IEEE 8-Bus Test System

The APS was tested on the IEEE 8-bus test system, as shown in Figure 6, for the original topological system, with an external grid linked in bus number 4. This system consists of 8 buses, which are connected with 7 lines and used 14 relays on the ends of the lines to protect these transmission lines. This system has two synchronous generators to feed 4 loads, in addition to the 400 MW for external grid entry and out-of-work [21,43].

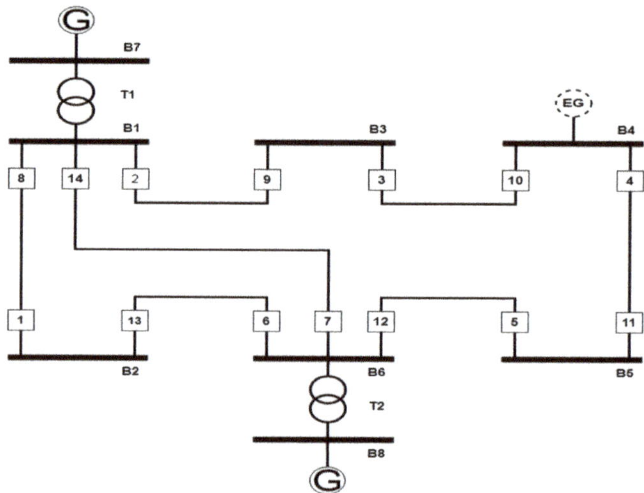

Figure 6. The single line diagram of IEEE 8-Bus.

This was a highly constrained, nonlinear optimization problem. It had 42 variables of design, which were tuned by optimization algorithms, in each case. The optimal solution was limited with minimum and maximum TDS, IP, TZ2, and T operate limiters. Additionally, they were constrained with a CTI value between the operation time of pairs primary and backup constraints. These constraints were 32 between DOCRs and equal to 40 between DOCRS and distance relays in the normal grid, while the external grid, on the state the constraints, became 34 between DOCRS and was still equal to 40 between the DOCRs and distance relays.

The ETAP program used to calculate three-phase fault currents is presented as Appendix A Table A1 for the normal grid. Additionally, for the second case, data was extracted from [43].

Table 1 lists the optimal settings for protection relays using both SBO and MSBO in normal topological settings, and the external grid is on. This table proved the MSBO has optimum solutions that are better than the optimum solution of SBO.

Table 1. IEEE 8-bus's relays setting.

Relay	Normal Topological Grid						With External DG					
	Original SBO			Modified SBO			Original SBO			Modified SBO		
	TDS	IP	TZ2	TDS	IP	TZ2	TDS	IP	TZ2	TDS	IP	TZ2
1	0.194	174.94	1.07	0.164	191.56	0.984	0.193	229.12	1.108	0.119	370.20	1.037
2	0.287	289.93	1	0.198	528.22	0.936	0.274	353.42	1.013	0.168	723.15	0.928
3	0.332	80.55	0.875	0.252	137.37	0.815	0.236	242.43	0.928	0.143	549.25	0.903
4	0.125	318.77	0.793	0.128	249.02	0.719	0.173	393.64	0.86	0.158	438.72	0.84
5	0.1	235.55	1.425	0.1	158.21	0.917	0.114	434.09	0.974	0.1	500.20	0.993
6	0.175	654.44	1.049	0.286	120.14	0.823	0.216	389.94	0.9	0.297	140.18	0.843
7	0.297	147.82	1.068	0.208	247.29	0.982	0.205	443.09	1.174	0.369	80.03	0.991
8	0.257	226.38	0.931	0.128	725.52	0.916	0.303	420.07	1.258	0.269	205.83	0.884
9	0.1	173.26	0.994	0.1	145.66	0.861	0.314	127.59	1.174	0.256	143.03	1.039
10	0.136	389.05	0.913	0.155	171.41	0.698	0.227	455.43	1.121	0.177	552.63	1.012
11	0.274	196.10	0.954	0.123	578.75	0.805	0.147	784.60	1.13	0.154	666.46	1.049
12	0.382	167.49	1.053	0.166	631.70	0.869	0.227	595.26	1.052	0.317	295.35	1.055
13	0.198	191.93	1.146	0.226	120.34	1.011	0.165	271.95	1.088	0.167	284.74	1.127
14	0.233	282.76	1.147	0.128	460.94	0.935	0.264	243.24	1.089	0.321	173.52	1.123
OF		33.705			28.072			35.388			32.601	

The normal case of the IEEE 8-bus test system constraints, which occur by optimum solution, was tabulated in Table 2. This table is for primary and backup operation time of DOCRs pairs relay in both near- and far-end. Addition to for constraints between DOCRs and distance relays. Table 3 has the same description as previous Table 2 but deal with another case in which the network is linked with the external grid. These tables show that modified algorithm satisfied all constraints.

Table 2. IEEE 8-bus's operation times of Relay's pairs in normal grid by MSBO.

Pair	Near-End						Far-End					
	DOCRs			D&DOCR			DOCRs			D&DOCR		
	T_p	T_b	CTI	T_p	T_{Z2B}	CTI	T_p	T_b	CTI	T_p	T_{Z2P}	CTI
1	0.423	0.623	0.200	0.423	0.823	0.400	0.784	1.027	0.243	0.784	0.984	0.2
2	0.582	0.784	0.202	0.582	0.984	0.402	0.736	2.390	1.655	0.736	0.936	0.2
3	0.582	0.782	0.200	0.582	0.982	0.400	0.736	1.674	0.939	0.736	0.936	0.2
4	0.535	0.736	0.200	0.535	0.936	0.400	0.615	0.943	0.329	0.615	0.815	0.2
5	0.399	0.615	0.216	0.399	0.815	0.416	0.519	0.752	0.233	0.519	0.719	0.2
6	0.318	0.519	0.201	0.318	0.719	0.401	0.717	1.738	1.021	0.717	0.917	0.2
7	0.517	0.717	0.201	0.517	0.917	0.401	0.623	—	—	0.623	0.823	0.2
8	0.517	0.735	0.218	0.517	0.935	0.418	0.623	—	—	0.623	0.823	0.2
9	0.497	0.717	0.220	0.497	0.917	0.420	0.782	—	—	0.782	0.982	0.2
10	0.497	0.811	0.314	0.497	1.011	0.514	0.782	—	—	0.782	0.982	0.2
11	0.457	0.782	0.325	0.457	0.982	0.525	0.716	—	—	0.716	0.916	0.2
12	0.457	0.661	0.203	0.457	0.861	0.403	0.716	—	—	0.716	0.916	0.2
13	0.298	0.498	0.200	0.298	0.698	0.400	0.661	1.216	0.555	0.661	0.861	0.2
14	0.405	0.605	0.200	0.405	0.805	0.400	0.498	0.924	0.426	0.498	0.698	0.2
15	0.469	0.669	0.200	0.469	0.869	0.400	0.605	0.877	0.272	0.605	0.805	0.2
16	0.532	0.811	0.280	0.532	1.011	0.480	0.669	1.471	0.802	0.669	0.869	0.2
17	0.532	0.735	0.203	0.532	0.935	0.403	0.669	2.480	1.811	0.669	0.869	0.2
18	0.504	0.716	0.212	0.504	0.916	0.412	0.811	7.976	7.165	0.811	1.011	0.2
19	0.394	0.784	0.389	0.394	0.984	0.589	0.735	—	—	0.735	0.935	0.2
20	0.394	0.661	0.266	0.394	0.861	0.466	0.735	—	—	0.735	0.935	0.2

Table 3. IEEE 8-bus's operation times of relay's pairs in case with extrnal grid by MSBO.

Pair	Near-End						Far-End					
	DOCRs			D&DOCR			DOCRs			D&DOCR		
	T_P	T_b	CTI	T_P	T_{Z2B}	CTI′	T_P	T_b	CTI	T_P	T_{Z2P}	CTI′
1	0.378	0.643	0.265	0.378	0.843	0.465	0.837	1.041	0.204	0.837	1.037	0.2
2	0.548	0.839	0.291	0.548	1.037	0.489	0.728	13.984	13.256	0.728	0.928	0.2
3	0.548	0.793	0.244	0.548	0.991	0.443	0.728	1.131	0.404	0.728	0.928	0.2
4	0.527	0.728	0.200	0.527	0.928	0.400	0.703	1.029	0.325	0.703	0.903	0.2
5	0.503	0.703	0.201	0.503	0.903	0.401	0.640	1.537	0.897	0.640	0.840	0.2
6	0.440	0.640	0.200	0.440	0.840	0.400	0.793	1.088	0.295	0.793	0.993	0.2
7	0.531	0.793	0.262	0.531	0.993	0.462	0.643	3.115	2.472	0.643	0.843	0.2
8	0.531	0.923	0.392	0.531	1.123	0.592	0.643	——	——	0.643	0.843	0.2
9	0.593	0.793	0.200	0.593	0.993	0.400	0.791	——	——	0.791	0.991	0.2
10	0.593	0.928	0.335	0.593	1.127	0.534	0.791	——	——	0.791	0.991	0.2
11	0.537	0.793	0.256	0.537	0.991	0.454	0.684	——	——	0.684	0.884	0.2
12	0.537	0.839	0.302	0.537	1.039	0.502	0.684	1.308	0.624	0.684	0.884	0.2
13	0.611	0.812	0.201	0.611	1.012	0.401	0.839	1.656	0.817	0.839	1.039	0.2
14	0.622	0.849	0.227	0.622	1.049	0.427	0.812	2.002	1.190	0.812	1.012	0.2
15	0.619	0.855	0.236	0.619	1.055	0.436	0.849	1.049	0.200	0.849	1.049	0.2
16	0.719	0.928	0.209	0.719	1.127	0.408	0.855	2.770	1.914	0.855	1.055	0.2
17	0.719	0.923	0.204	0.719	1.123	0.404	0.855	1.423	0.568	0.855	1.055	0.2
18	0.485	0.685	0.200	0.485	0.884	0.399	0.927	1.182	0.255	0.927	1.127	0.2
19	0.639	0.839	0.200	0.639	1.037	0.398	0.923	——	——	0.923	1.123	0.2
20	0.639	0.839	0.200	0.639	1.039	0.400	0.923	——	——	0.923	1.123	0.2

The convergence characteristics curves of SBO and MSBO for the normal case and the other case are presented in Figures 7 and 8, respectively. And the penalty occurred by SBO and MSBO during running the optimum algorithm shown in Figure 9. For the normal case while the state of the external grid is shown in Figure 10. These figures showed the convergence of MSBO is better and faster than the original SBO convergence. And the ability of MSBO to avoid penalty and pass constraints quickly.

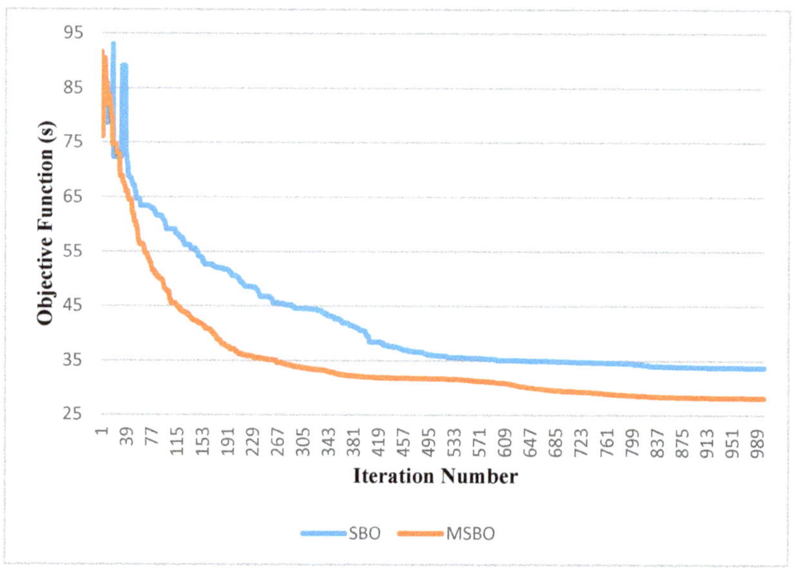

Figure 7. Convergence characteristics of SBO and MSBO in normal case of IEEE 8-bus.

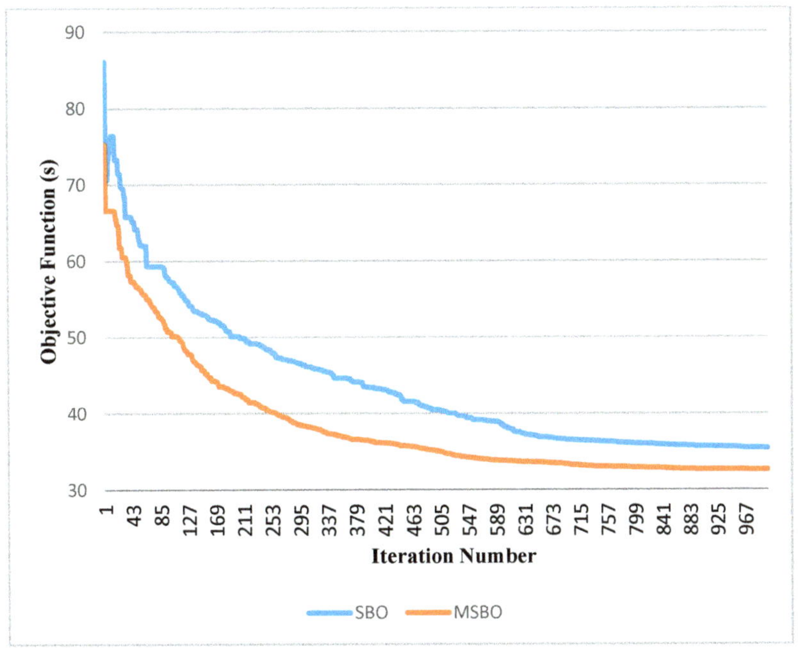

Figure 8. Convergence characteristics of SBO and MSBO in external grid on case of IEEE 8-bus.

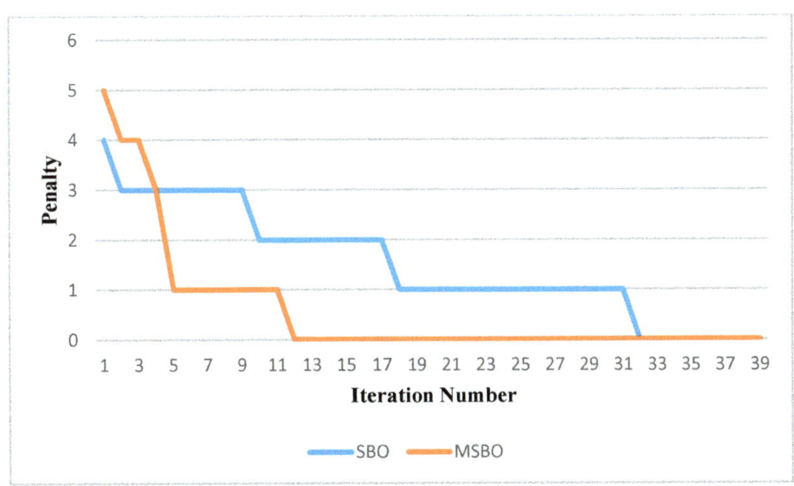

Figure 9. Penalty between SBO and MSBO of IEEE 8-bus test system normal case.

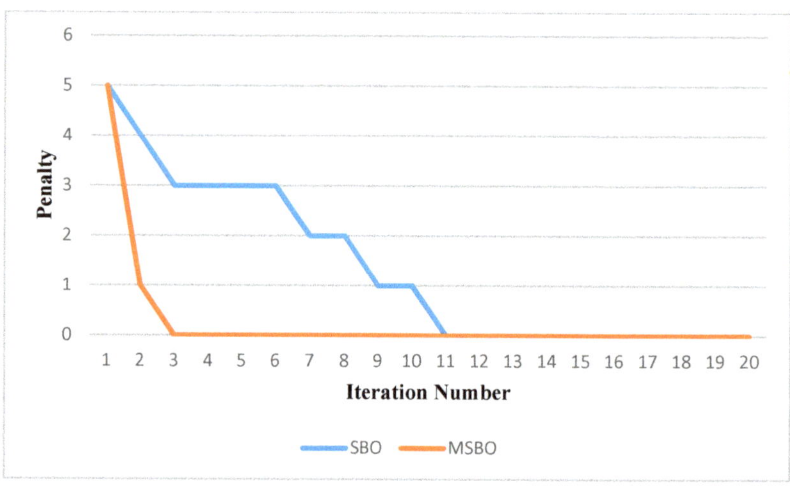

Figure 10. Penalty between SBO and MSBO of IEEE 8-bus test system with external grid on work.

In the normal case, the original SBO has OF with value 33.705 s while MSBO has OF with value 28.072 s and MSBO has 33.705 s after 300 iterations and MSBO passed penalty after 12 iterations while original SBO continuous to iteration 32 to pass system's constraints. For another case, the external grid stat on operation OF becomes 32.601 s and 35.388 s for MSBO and SBO respectively. MSBO reached 35.350 s after 477 iterations. The penalty passed after 11 iterations in the case of SBO while MSBO passed after three iterations. All of these prove the ability of MSBO to increase its exploration and exploitation more than the original algorithm.

4.2. Test System II: IEEE 14-Bus Distribution Network

The IEEE 14-bus distribution network which is shown in Figure 11. Which is a downstream section of IEEE 14-bus. This distribution network has two distribution transformers connected at buses number 1 and 2 to supply it. Each transmission line has a protection relay at every end of the line to form 16 relays [38,44].

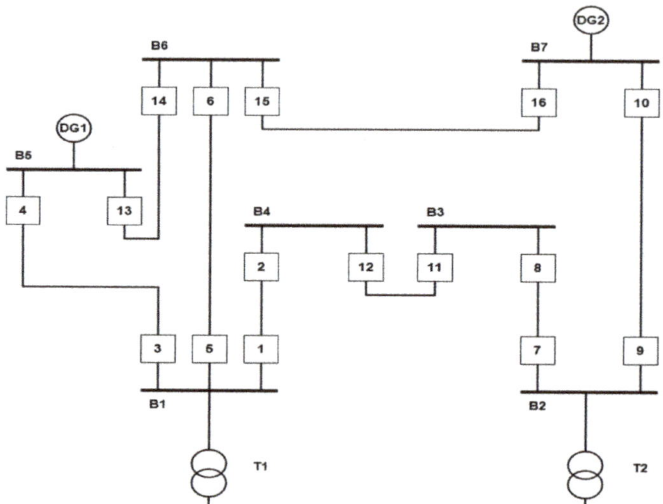

Figure 11. The single line diagram for 14 bus distribution network.

Test system modified with addition 2 DGs connected at buses number 5 and 7 with power equals 5MVA and power factor nominally is 0.9 lagging and their type are synchronous. This modification is from [45] and 3 phase short circuit currents at both near-end and far-end from [38].

The current transformer's ratios of relays are 120, 120, 120, 40, 120, 40, 120, 120, 120, 80, 80, 80, 40, 40, 80, and 80, for relays from 1 to 16, respectively [38].

In this test system, optimization algorithms tuned 48 variables that are limited by minimum and maximum values. Then constrained between operation time values of relays by CTI. This test system is more constrained than the previous test system by 41 between primary and backup DOCRs and 44 between DOCRs and distance relays, in both ends (near and far).

Optimal settings tuned by SBO and MSBO algorithms. Those were tabulated in Table 4 for the normal case and another case. These optimization solutions passed CTI constraints between relays pairs in normal case in Table 5. Then, Table 6 proved the ability of optimization solutions to pass the CTI constraint between relays pairs in DGs working case.

Figure 12 shows the convergence characteristics curve of both SBO and MSBO algorithms in the normal case while in another case after DGs work is shown in Figure 13. During the tuning process for relays setting by optimization algorithms SBO and MSBO, the penalty of both are shown in Figures 14 and 15 for the normal case and the other DGs case, respectively.

In the normal case, MSBO has OF with a value of 34.806 s and is better than SBO with 2.05 s. MSBO reached 36.860 s faster than SBO by 328 iterations and passed penalty after 22 iterations the original SBO passed after 71 iterations. In another case, MSBO reaches SBO's OF after 482 iterations and at the end of the run reaches 51.068 s as OF and better than SBO by 6.2 s. MSBO passed constraints penalty after 113 iterations while SBO still penalty to 183 iterations.

Table 4. IEEE 14-bus distribution network's relays setting.

Relay	Normal Topological						With External DG					
	Original SBO			Modified SBO			Original SBO			Modified SBO		
	TDS	IP	TZ2	TDS	IP	TZ2	TDS	IP	TZ2	TDS	IP	TZ2
1	0.223	235.93	1.246	0.214	157.86	0.984	0.386	392.93	1.549	0.384	337.37	1.442
2	0.11	185.11	0.885	0.159	149.46	1.028	0.37	188.60	1.455	0.209	314.56	1.153
3	0.135	232.94	0.931	0.155	164.17	0.858	0.434	139.04	1.255	0.241	465.02	1.243
4	0.156	51.91	1.103	0.232	20.02	0.937	0.376	52.32	1.455	0.402	39.28	1.376
5	0.243	80.15	0.824	0.211	142.21	0.895	0.492	231.73	1.514	0.289	394.61	1.18
6	0.361	28.91	1.069	0.219	63.82	0.937	0.356	131.99	1.454	0.621	28.58	1.407
7	0.241	140.62	0.865	0.164	425.78	1.025	0.731	71.12	1.446	0.33	361.09	1.172
8	0.258	60.58	0.953	0.15	117.13	0.815	0.447	106.48	1.367	0.283	233.98	1.266
9	0.105	308.66	0.864	0.162	197.06	0.923	0.349	347.18	1.423	0.363	260.94	1.308
10	0.121	187.97	1.162	0.116	167.35	1.014	0.311	239.60	1.566	0.447	72.25	1.299
11	0.158	176.02	0.854	0.282	85.18	1.008	0.529	119.67	1.414	0.324	209.62	1.121
12	0.229	150.99	1.294	0.165	133.91	0.93	0.534	105.85	1.499	0.569	69.66	1.398
13	0.151	101.59	1.077	0.185	58.68	0.929	0.399	85.44	1.345	0.464	60.89	1.362
14	0.205	65.11	0.8	0.313	27.29	0.861	0.46	54.40	1.088	0.444	58.67	1.078
15	0.136	179.69	0.953	0.139	152.58	0.878	0.489	106.17	1.38	0.298	276.49	1.297
16	0.119	201.72	1.092	0.198	89.52	0.984	0.491	102.67	1.494	0.4	141.69	1.408
OF		36.86			34.806			57.268			51.068	

Table 5. IEEE 14-bus distribution network's operation times of Relay's pairs in normal case by MSBO.

Pair	Near-End						Far-End					
	DOCRs			D&DOCR			DOCRs			D&DOCR		
	T_p	T_b	CTI	T_p	T_{Z2B}	CTI′	T_p	T_b	CTI	T_p	T_{Z2P}	CTI′
1	0.537	0.737	0.201	0.537	0.937	0.401	0.784	1.096	0.312	0.784	0.984	0.2
2	0.537	0.737	0.201	0.537	0.937	0.401	0.784	1.272	0.488	0.784	0.984	0.2
3	0.575	0.808	0.233	0.575	1.008	0.433	0.828	1.032	0.203	0.828	1.028	0.2
4	0.374	0.828	0.455	0.374	1.028	0.655	0.658	1.997	1.339	0.658	0.858	0.2
5	0.374	0.737	0.364	0.374	0.937	0.564	0.658	——	——	0.658	0.858	0.2
6	0.427	0.661	0.234	0.427	0.861	0.434	0.737	1.907	1.169	0.737	0.937	0.2
7	0.504	0.828	0.325	0.504	1.028	0.525	0.695	1.352	0.657	0.695	0.895	0.2
8	0.504	0.737	0.234	0.504	0.937	0.434	0.695	——	——	0.695	0.895	0.2
9	0.524	0.729	0.204	0.524	0.929	0.404	0.737	——	——	0.737	0.937	0.2
10	0.524	0.784	0.259	0.524	0.984	0.459	0.737	0.938	0.201	0.737	0.937	0.2
11	0.613	0.814	0.2	0.613	1.014	0.4	0.825	1.546	0.721	0.825	1.025	0.2
12	0.529	0.73	0.201	0.529	0.93	0.401	0.615	0.856	0.241	0.615	0.815	0.2
13	0.413	0.615	0.202	0.413	0.815	0.402	0.723	7.393	6.669	0.723	0.923	0.2
14	0.476	0.676	0.2	0.476	0.878	0.403	0.814	1.703	0.89	0.814	1.014	0.2
15	0.621	0.825	0.204	0.621	1.025	0.404	0.808	1.604	0.796	0.808	1.008	0.2
16	0.54	0.784	0.244	0.54	0.984	0.444	0.73	1.066	0.336	0.73	0.93	0.2
17	0.452	0.658	0.205	0.452	0.858	0.405	0.729	2.547	1.819	0.729	0.929	0.2
18	0.495	0.695	0.2	0.495	0.895	0.4	0.661	2.583	1.922	0.661	0.861	0.2
19	0.495	0.784	0.289	0.495	0.984	0.489	0.661	1.187	0.526	0.661	0.861	0.2
20	0.391	0.695	0.304	0.391	0.895	0.504	0.678	1.445	0.766	0.678	0.878	0.2
21	0.391	0.729	0.337	0.391	0.929	0.537	0.678	1.66	0.982	0.678	0.878	0.2
22	0.522	0.723	0.201	0.522	0.923	0.401	0.784	1.886	1.102	0.784	0.984	0.2

Table 6. IEEE 14-bus distribution network's operation times of relays pairs with DGs by MSBO.

Pair	Near-End						Far-End					
	DOCRs			D&DOCR			DOCRs			D&DOCR		
	T_p	T_b	CTI	T_p	T_{Z2B}	CTI	T_p	T_b	CTI	T_p	T_{Z2P}	CTI
1	0.973	1.18	0.207	0.973	1.376	0.403	1.242	1.871	0.629	1.242	1.442	0.2
2	0.973	1.205	0.232	0.973	1.407	0.434	1.242	2.081	0.839	1.242	1.442	0.2
3	0.718	0.921	0.203	0.718	1.121	0.403	0.953	1.159	0.207	0.953	1.153	0.2
4	0.644	0.956	0.312	0.644	1.153	0.509	1.043	1.508	0.464	1.043	1.243	0.2
5	0.644	1.205	0.562	0.644	1.407	0.763	1.043	——	——	1.043	1.243	0.2
6	0.677	0.881	0.204	0.677	1.078	0.401	1.176	26.389	25.213	1.176	1.376	0.2
7	0.752	0.956	0.204	0.752	1.153	0.401	0.98	1.506	0.525	0.98	1.18	0.2
8	0.752	1.18	0.428	0.752	1.376	0.624	0.98	——	——	0.98	1.18	0.2
9	0.948	1.162	0.214	0.948	1.362	0.414	1.207	——	——	1.207	1.407	0.2
10	0.948	1.206	0.258	0.948	1.408	0.46	1.207	1.409	0.202	1.207	1.407	0.2
11	0.873	1.1	0.227	0.873	1.299	0.427	0.972	1.323	0.351	0.972	1.172	0.2
12	0.947	1.181	0.234	0.947	1.398	0.451	1.066	1.271	0.205	1.066	1.266	0.2
13	0.832	1.059	0.227	0.832	1.266	0.433	1.108	3.803	2.695	1.108	1.308	0.2
14	0.892	1.097	0.205	0.892	1.297	0.405	1.099	1.821	0.722	1.099	1.299	0.2
15	0.768	0.972	0.204	0.768	1.172	0.404	0.921	1.22	0.299	0.921	1.121	0.2
16	1.038	1.242	0.204	1.038	1.442	0.404	1.198	1.611	0.413	1.198	1.398	0.2
17	0.826	1.032	0.206	0.826	1.243	0.417	1.162	6.046	4.884	1.162	1.362	0.2
18	0.691	0.98	0.29	0.691	1.18	0.49	0.878	2.43	1.553	0.878	1.078	0.2
19	0.691	1.206	0.515	0.691	1.408	0.717	0.878	1.489	0.612	0.878	1.078	0.2
20	0.766	0.98	0.215	0.766	1.18	0.415	1.097	1.637	0.54	1.097	1.297	0.2
21	0.766	1.162	0.397	0.766	1.362	0.597	1.097	1.603	0.506	1.097	1.297	0.2
22	0.908	1.108	0.2	0.908	1.308	0.4	1.208	1.735	0.527	1.208	1.408	0.2

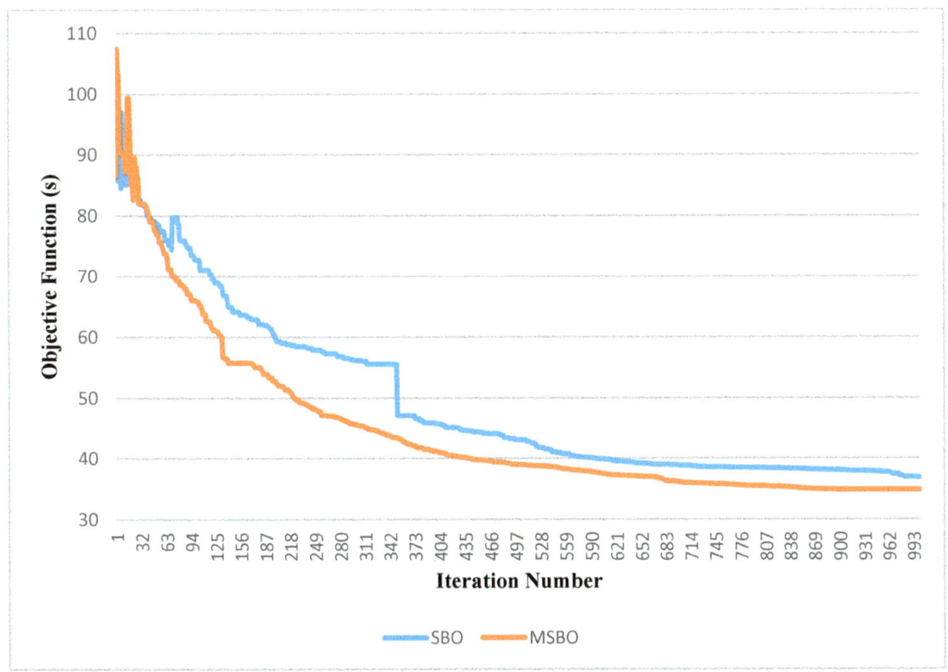

Figure 12. Convergence characteristics of SBO and MSBO in normal case of IEEE 14-bus distribution network.

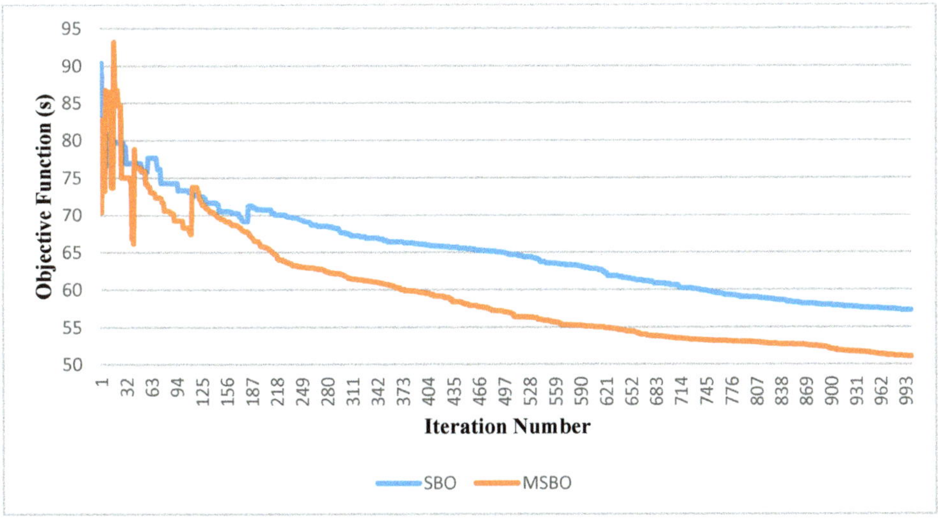

Figure 13. Convergence characteristics of SBO and MSBO with DGs case of IEEE 14-bus distribution network.

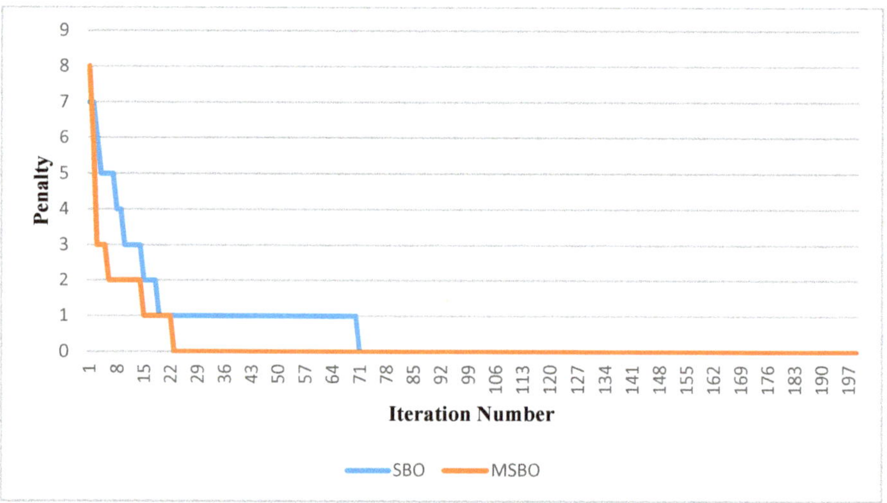

Figure 14. Penalty between SBO and MSBO of IEEE 14-bus distribution network normal case.

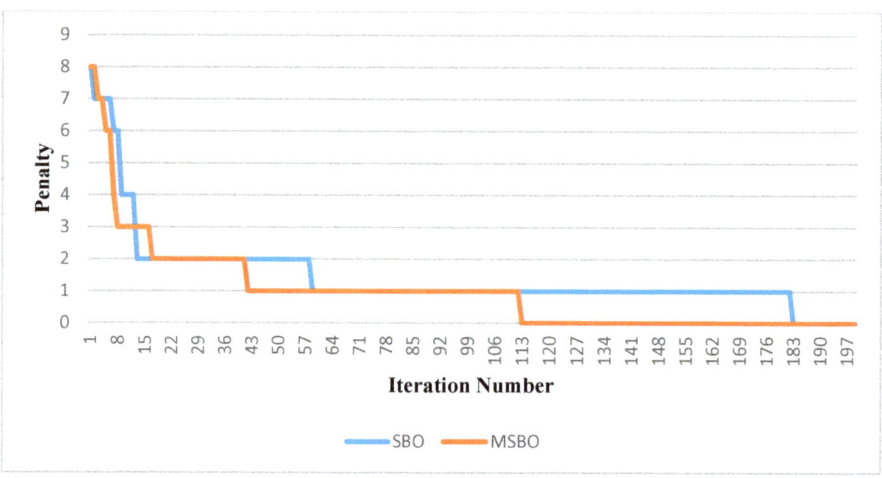

Figure 15. Penalty between SBO and MSBO of IEEE 14-bus distribution network with DGs.

4.3. Verification of MSBO Using Etap 12.6.0

Etap is used to verify the results obtained by the proposed algorithm MSBO on 8 bus normal grid, three-phase fault at transmission line between third and fourth bus-bars. This fault applied in both the near end and far end. As simulation by Etap as shown in Figure 16. Relay 3 is the primary relay that operates at 0.535 s and 0.615 s in the near and far end, respectively. While relay 2 is its backup relay which operates at 0.736 s and 0.943 in near and far ends, respectively. And these verify the CTI is more than or equal to 0.2 s and DOCRs without miscoordination. That simulation is also done at the transmission line between the fifth and sixth busbars additional to relays 5 as primary relay and 4 as backup re-lay, the operation time at both ends near and far of this pair relay as 0.318 s, 0.519 s, 0.717 s, and 1.738 s, respectively. The simulation is presented in Figure 17 proved that there is no miscoordination between DOCRs.

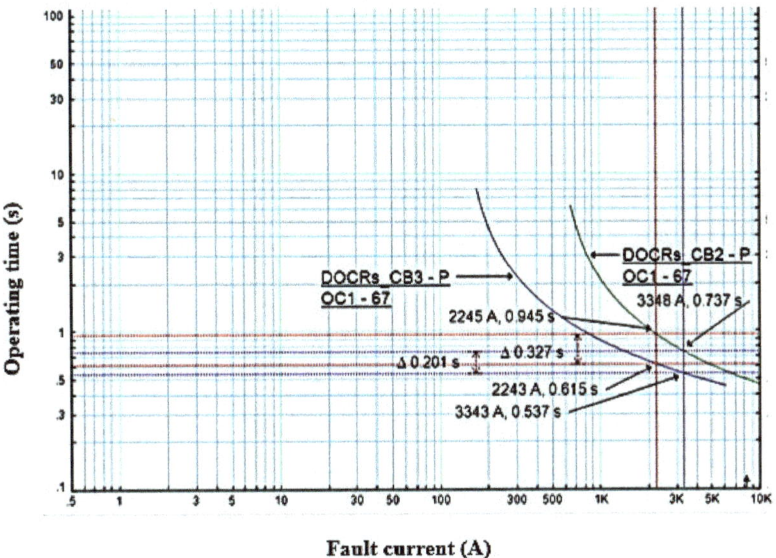

Figure 16. Operating times for relays 3 and 2.

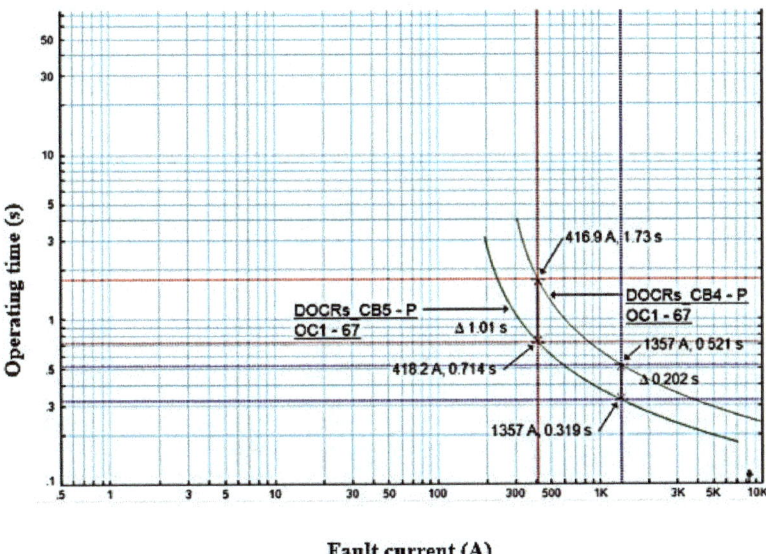

Figure 17. Operating times for relays 5 and 4.

Finally, simulation is done at the transmission line which is connected between the first and third busbars. It is noticed from Figure 18. Operating times at the near end for primary (relay 9), and backup (relay 10) are 0.298 s, and 0.498 s, respectively. At the far end the operating time of primary and backup relays are 0.661 s, and 1.216 s, respectively. This case avoids miscoordination too.

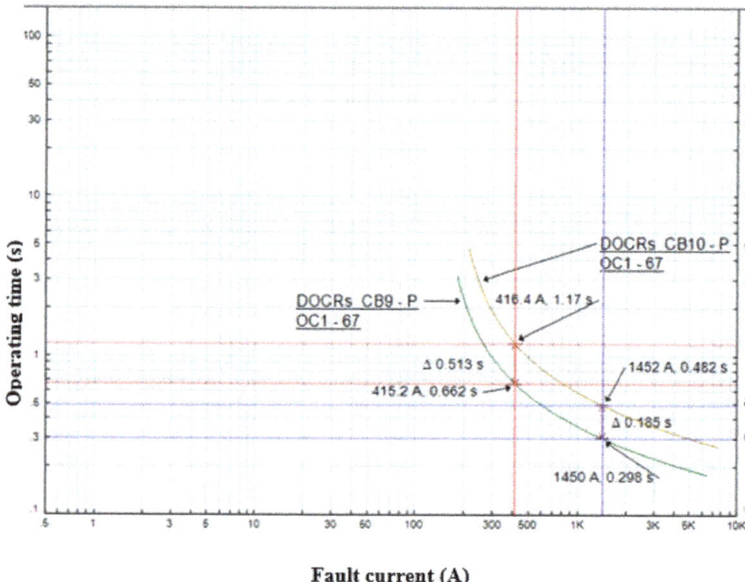

Figure 18. Operating times for relays 9 and 10.

5. Conclusions

In this research work, the APS passed the miscoordination problem between primary and backup DOCRs, as well as miscoordination between DOCRs and distance relays. The results demonstrated that APS has the potential to coordinate protection relays with appropriate settings to face the problem of DGs in the distribution network. Without miscoordination between protection relays, the power system can investigate the advantages of both distance relays and DOCRs. In addition to the role of the modified optimization algorithm, which improved the optimal values and reduced operation relay's time, all operation times of both primary DOCRs and distance second zone's time were set below the maximum operation time. The modified algorithm has better convergence characteristics and achieves better optimal values with fewer iterations.

Author Contributions: Conceptualization, S.K. and A.K.; Data curation, S.K. and A.K.; formal analysis, S.K., A.K., M.T.-V., F.A.B. and M.I.M.; funding acquisition, F.A.B. and M.I.M.; investigation, M.A., S.K., A.K. and M.T.-V.; methodology, M.A., S.K. and A.K.; project administration, S.K., M.T.-V., F.A.B. and M.I.M.; resources, S.K. and M.T.-V.; software, S.K.; supervision, S.K., A.K., M.T.-V., F.A.B. and M.I.M.; validation, M.A., S.K. and A.K.; visualization, M.T.-V., F.A.B. and M.I.M.; writing—original draft, M.A. and A.K.; writing—review & editing, S.K. and M.T.-V. All authors have read and agreed to the published version of the manuscript.

Funding: The authors thank the support of the National Research and Development Agency of Chile (ANID), ANID/Fondap/15110019.

Institutional Review Board Statement: Not applicable.

Informed Consent Statement: Not applicable.

Data Availability Statement: Not applicable.

Conflicts of Interest: The authors declare no conflict of interest.

Appendix A

Table A1. Three-phase short circuit currents for IEEE 8-bus in the normal case by ETAP in both near end and far end.

Pair	Primary Relay	Back-Up Relay	Near-End		Far-End	
			Primary	Back-Up	Primary	Back-Up
1	1	6	2703	2703	812	812
2	2	1	5390	812	3347	309
3	2	7	5390	1540	3347	586
4	3	2	3347	3347	2243	2243
5	4	3	2243	2243	1361	1361
6	5	4	1361	1361	416	416
7	6	5	4995	416	2703	20
8	6	14	4995	1540	2703	72
9	7	5	4267	416	1540	REV
10	7	13	4267	812	1540	REV
11	8	7	4995	1540	2507	REV
12	8	9	4995	416	2507	REV
13	9	10	1453	1453	416	416
14	10	11	2344	2344	1453	1453
15	11	12	3495	3495	2344	2344
16	12	13	5390	812	3495	348
17	12	14	5390	1540	3495	661
18	13	8	2507	2507	812	812
19	14	1	4267	812	1540	REV
20	14	9	4267	416	1540	REV

References

1. Abdelhamid, M.; Kamel, S.; Mohamed, M.A.; Aljohani, M.; Rahmann, C.; Mosaad, M.I. Political Optimization Algorithm for Optimal Coordination of Directional Overcurrent Relays. In Proceedings of the 2020 IEEE Electric Power and Energy Conference (EPEC), Edmonton, AB, Canada, 9–10 November 2020; pp. 1–7. [CrossRef]
2. Damchi, Y.; Sadeh, J.; Mashhadi, H.R. Considering pilot protection in the optimal coordination of distance and directional overcurrent relays. *Iran. J. Electr. Electron. Eng.* **2015**, *11*, 154–164.
3. Perez, L.G.; Urdaneta, A.J. Optimal computation of distance relays second zone timing in a mixed protection scheme with directional overcurrent relays. *IEEE Trans. Power Deliv.* **2001**, *16*, 385–388. [CrossRef]
4. Khederzadeh, M. Back-up protection of distance relay second zone by directional overcurrent relays with combined curves. In Proceedings of the 2006 IEEE Power Engineering Society General Meeting, Montreal, QC, Canada, 18–22 June 2006; p. 6. [CrossRef]
5. Sarwagya, K.; Nayak, P.K.; Ranjan, S. Optimal coordination of directional overcurrent relays in complex distribution networks using sine cosine algorithm. *Electr. Power Syst. Res.* **2020**, *187*, 106435. [CrossRef]
6. Ates, Y.; Uzunoglu, M.; Karakas, A.; Boynuegri, A.R.; Nadar, A.; Dag, B. Implementation of adaptive relay coordination in distribution systems including distributed generation. *J. Clean. Prod.* **2016**, *112*, 2697–2705. [CrossRef]
7. Akdag, O.; Yeroglu, C. Optimal directional overcurrent relay coordination using MRFO algorithm: A case study of adaptive protection of the distribution network of the Hatay province of Turkey. *Electr. Power Syst. Res.* **2021**, *192*, 106998. [CrossRef]
8. Mansour, M.M.; Mekhamer, S.F.; El-Kharbawe, N. A modified particle swarm optimizer for the coordination of directional overcurrent relays. *IEEE Trans. Power Deliv.* **2007**, *22*, 1400–1410. [CrossRef]
9. Noghabi, A.S.; Sadeh, J.; Mashhadi, H.R. Considering different network topologies in optimal overcurrent relay coordination using a hybrid GA. *IEEE Trans. Power Deliv.* **2009**, *24*, 1857–1863. [CrossRef]
10. Albasri, F.A.; Alroomi, A.R.; Talaq, J.H. Optimal coordination of directional overcurrent relays using biogeography-based optimization algorithms. *IEEE Trans. Power Deliv.* **2015**, *30*, 1810–1820. [CrossRef]
11. Shih, M.Y.; Enriquez, A.C. Novel coordination of DOCRs using trigonometric differential evolution algorithm. *IEEE Lat. Am. Trans.* **2015**, *13*, 1605–1611. [CrossRef]
12. Khurshaid, T.; Wadood, A.; Farkoush, S.G.; Kim, C.-H.; Yu, J.; Rhee, S.-B. Improved firefly algorithm for the optimal coordination of directional overcurrent relays. *IEEE Access* **2019**, *7*, 78503–78514. [CrossRef]
13. Khurshaid, T.; Wadood, A.; Farkoush, S.G.; Yu, J.; Kim, C.-H.; Rhee, S.-B. An improved optimal solution for the directional overcurrent relays coordination using hybridized whale optimization algorithm in complex power systems. *IEEE Access* **2019**, *7*, 90418–90435. [CrossRef]

14. Yu, J.; Kim, C.-H.; Rhee, S.-B. Oppositional jaya algorithm with distance-adaptive coefficient in solving directional over current relays coordination problem. *IEEE Access* **2019**, *7*, 150729–150742. [CrossRef]
15. Korashy, A.; Kamel, S.; Alquthami, T.; Jurado, F. Optimal coordination of standard and non-standard direction overcurrent relays using an improved moth-flame optimization. *IEEE Access* **2020**, *8*, 87378–87392. [CrossRef]
16. Abdelhamid, M.; Kamel, S.; Mohamed, M.A.; Rahmann, C. An Effective Approach for Optimal Coordination of Directional Overcurrent Relays Based on Artificial Ecosystem Optimizer. In Proceedings of the 2021 IEEE International Conference on Automation/XXIV Congress of the Chilean Association of Automatic Control (ICA-ACCA), Valparaíso, Chile, 22–26 March 2021; pp. 1–6. [CrossRef]
17. Korashy, A.; Kamel, S.; Houssein, E.H.; Jurado, F.; Hashim, F.A. Development and application of evaporation rate water cycle algorithm for optimal coordination of directional overcurrent relays. *Expert Syst. Appl.* **2021**, *185*, 115538. [CrossRef]
18. Marcolino, M.H.; Leite, J.B.; Mantovani, J.R.S. Optimal coordination of overcurrent directional and distance relays in meshed networks using genetic algorithm. *IEEE Lat. Am. Trans.* **2015**, *13*, 2975–2982. [CrossRef]
19. Tiwari, R.; Singh, R.K.; Choudhary, N.K. A Comparative Analysis of Optimal Coordination of Distance and Overcurrent Relays with Standard Relay Characteristics using GA, GWO and WCA. In Proceedings of the 2020 IEEE Students Conference on Engineering & Systems (SCES), Prayagraj, India, 10–12 July 2020; pp. 1–6.
20. Bangar, P.A.; Kalage, A.A. Optimum coordination of overcurrent and distance relays using JAYA optimization algorithm. In Proceedings of the 2017 International Conference on Nascent Technologies in Engineering (ICNTE), Vashi, India, 27–28 January 2017; pp. 1–5.
21. Rivas, A.E.L.; Pareja, L.A.G.; Abrão, T. Coordination of distance and directional overcurrent relays using an extended continuous domain ACO algorithm and an hybrid ACO algorithm. *Electr. Power Syst. Res.* **2019**, *170*, 259–272. [CrossRef]
22. Vijayakumar, D.; Nema, R.K. A novel optimal setting for directional over current relay coordination using particle swarm optimization. *Int. J. Electr. Power Energy Syst. Eng.* **2008**, *1*, 82.
23. Singh, D.K.; Gupta, S. Optimal coordination of directional overcurrent relays: A genetic algorithm approach. In Proceedings of the 2012 IEEE Students' Conference on Electrical, Electronics and Computer Science, Bhopal, India, 1–2 March 2012; pp. 1–4.
24. Moirangthem, J.; Krishnanand, K.R.; Dash, S.S.; Ramaswami, R. Adaptive differential evolution algorithm for solving non-linear coordination problem of directional overcurrent relays. *IET Gener. Transm. Distrib.* **2013**, *7*, 329–336. [CrossRef]
25. Shih, M.Y.; Salazar, C.A.C.; Enríquez, A.C. Adaptive directional overcurrent relay coordination using ant colony optimization. *IET Gener. Transm. Distrib.* **2015**, *9*, 2040–2049. [CrossRef]
26. Chawla, A.; Bhalja, B.R.; Panigrahi, B.K.; Singh, M. Gravitational search based algorithm for optimal coordination of directional overcurrent relays using user defined characteristic. *Electr. Power Compon. Syst.* **2018**, *46*, 43–55. [CrossRef]
27. Tjahjono, A.; Anggriawan, D.O.; Faizin, A.K.; Priyadi, A.; Pujiantara, M.; Taufik, T.; Purnomo, M.H. Adaptive modified firefly algorithm for optimal coordination of overcurrent relays. *IET Gener. Transm. Distrib.* **2017**, *11*, 2575–2585. [CrossRef]
28. ElSayed, S.K.; Elattar, E.E. Hybrid Harris hawks optimization with sequential quadratic programming for optimal coordination of directional overcurrent relays incorporating distributed generation. *Alex. Eng. J.* **2021**, *60*, 2421–2433. [CrossRef]
29. Farshchin, M.; Maniat, M.; Camp, C.V.; Pezeshk, S. School based optimization algorithm for design of steel frames. *Eng. Struct.* **2018**, *171*, 326–335. [CrossRef]
30. Rao, R.V.; Savsani, V.J.; Vakharia, D.P. Teaching–learning-based optimization: A novel method for constrained mechanical design optimization problems. *Comput. Aided Des.* **2011**, *43*, 303–315. [CrossRef]
31. Degertekin, S.O.; Tutar, H.; Lamberti, L. School-based optimization for performance-based optimum seismic design of steel frames. *Eng. Comput.* **2021**, *37*, 3283–3297. [CrossRef]
32. Abdelghany, R.Y.; Kamel, S.; Ramadan, A.; Sultan, H.M.; Rahmann, C. Solar Cell Parameter Estimation Using School-Based Optimization Algorithm. In Proceedings of the 2021 IEEE International Conference on Automation/XXIV Congress of the Chilean Association of Automatic Control (ICA-ACCA), Valparaíso, Chile, 22–26 March 2021; pp. 1–6. [CrossRef]
33. Sampaio, F.C.; Leão, R.P.S.; Sampaio, R.F.; Melo, L.S.; Barroso, G.C. A multi-agent-based integrated self-healing and adaptive protection system for power distribution systems with distributed generation. *Electr. Power Syst. Res.* **2020**, *188*, 106525. [CrossRef]
34. Reis, F.B.d.; Pinto, J.O.C.P.; Reis, F.S.d.; Issicaba, D.; Rolim, J.G. Multi-agent dual strategy based adaptive protection for microgrids. *Sustain. Energy Grids Netw.* **2021**, *27*, 100501. [CrossRef]
35. Cui, Q.; Weng, Y. An environment-adaptive protection scheme with long-term reward for distribution networks. *Int. J. Electr. Power Energy Syst.* **2020**, *124*, 106350. [CrossRef]
36. Elmitwally, A.; Kandil, M.S.; Gouda, E.; Amer, A. Mitigation of DGs impact on variable-topology meshed network protection system by optimal fault current limiters considering overcurrent relay coordination. *Electr. Power Syst. Res.* **2020**, *186*, 106417. [CrossRef]
37. Yazdaninejadi, A.; Nazarpour, D.; Golshannavaz, S. Sustainable electrification in critical infrastructure: Variable characteristics for overcurrent protection considering DG stability. *Sustain. Cities Soc.* **2020**, *54*, 102022. [CrossRef]
38. Rajput, V.N.; Adelnia, F.; Pandya, K.S. Optimal coordination of directional overcurrent relays using improved mathematical formulation. *IET Gener. Transm. Distrib.* **2018**, *12*, 2086–2094. [CrossRef]
39. Moravej, Z.; Ooreh, O.S. Coordination of distance and directional overcurrent relays using a new algorithm: Grey wolf optimizer. *Turk. J. Electr. Eng. Comput. Sci.* **2018**, *26*, 3130–3144. [CrossRef]

40. Mohammadi, R.; Abyaneh, H.A.; Rudsari, H.M.; Fathi, S.H.; Rastegar, H. Overcurrent relays coordination considering the priority of constraints. *IEEE Trans. Power Deliv.* **2011**, *26*, 1927–1938. [CrossRef]
41. Shih, M.Y.; Conde, A.; Ángeles-Camacho, C.; Fernández, E.; Leonowicz, Z.; Lezama, F.; Chan, J. A two stage fault current limiter and directional overcurrent relay optimization for adaptive protection resetting using differential evolution multi-objective algorithm in presence of distributed generation. *Electr. Power Syst. Res.* **2021**, *190*, 106844. [CrossRef]
42. Camp, C.V.; Farshchin, M. Design of space trusses using modified teaching-learning based optimization. *Eng. Struct.* **2014**, *62*, 87–97. [CrossRef]
43. Yang, M.-T.; Liu, A. Applying hybrid PSO to optimize directional overcurrent relay coordination in variable network topologies. *J. Appl. Math.* **2013**, *2013*, 879078. [CrossRef]
44. Christie, R. Power systems test case archive. *Electr. Eng. Dept. Univ. Wash.* **2000**, *108*. Available online: https://labs.ece.uw.edu/pstca/ (accessed on 10 October 2021).
45. Saleh, K.A.; Zeineldin, H.H.; Al-Hinai, A.; El-Saadany, E.F. Optimal coordination of directional overcurrent relays using a new time-current-voltage characteristic. *IEEE Trans. Power Deliv.* **2014**, *30*, 537–544. [CrossRef]

 electronics

MDPI

Article

Uncertainty Costs Optimization of Residential Solar Generators Considering Intraday Markets

Julian Garcia-Guarin, David Alvarez and Sergio Rivera *

Electrical Engineering, Universidad Nacional de Colombia, Bogotá 110111, Colombia; pjgarciag@unal.edu.co (J.G.-G.); dlalvareza@unal.edu.co (D.A.)
* Correspondence: srriverar@unal.edu.co; Tel.: +56-3204632806

Abstract: The uncertainty of solar generation and the bull market are unavoidable in energy dispatch. The purpose of this research is to validate an uncertainty cost function of residential photovoltaic energy in a real microgrid by varying the number of auctions in intraday markets. Therefore, the following procedure is proposed. First, the variability of photovoltaic generation is quantified through Monte Carlo simulations. Second, a statistical function calculates the variability costs of photovoltaic generation. Third, the uncertainty costs are estimated by varying intraday auction markets. Other complementary services are added to the network, such as battery exchange stations for electric vehicles, demand response loads, market power restrictions, and energy storage systems, which are estimated as total costs in an index ranking. The total costs are optimized in a benchmark microgrid and take complimentary services as a black box. Only the uncertainty costs of residential solar generators are discriminated. The main findings were that (1) the uncertainty costs have an error of less than 0.0168% compared to the Monte Carlo simulations and that (2) the uncertainty costs of solar generation are reduced with a decreasing trend to a more significant number of auction markets in intraday markets.

Keywords: electric markets; photovoltaic generation; Monte Carlo simulations

Citation: Garcia-Guarin, J.; Alvarez, D.; Rivera, S. Uncertainty Costs Optimization of Residential Solar Generators Considering Intraday Markets. *Electronics* **2021**, *10*, 2826. https://doi.org/10.3390/electronics10222826

Academic Editors: Nebojša Raičević, Vasilija Šarac and Marinko Barukčić

Received: 13 October 2021
Accepted: 15 November 2021
Published: 17 November 2021

1. Introduction

The greater diffusion of renewable energies mitigates the environmental deterioration caused by greenhouse gases due to conventional electricity generation (Gen) [1]. Power microgrids (MGs)are complex because they face uncertainties such as demand forecasts, electric vehicles (EVs), battery swapping stations (BSSs), market price (MP) variability, and renewable energy forecasts [2–7]. Gen is strongly influenced by the variability of electricity market (EM) prices, which seek to minimize operating costs using energy sources such as the solar power [8]. Furthermore, pivotal agents and monopolies should be reduced because they produce market power. Traditionally, the Herfindahl Hirschman index and the Residual Supply Index have been used to monitor EMs [9].

The prediction of photovoltaic (PV) energy has been extensively studied with Monte Carlo simulations [8]. However, the lack of reliable information on solar Gen makes energy delivery less efficient. Weather conditions such as unpredictable winds prevent forecasts from being accurate. In such a scenario, uncertainty is inherent and cannot be eliminated in planning [8]. Additionally, energy is available in Ems, where agents can buy and sell power [10]. Intraday markets (IMs) present an additional complexity that should be responsible for mismatches in scheduling on the day of operation. These imbalances are produced by changes in the forecasts of the load or PV Gen [11].

The regularization of the electric generators that are involved in energy dispatch must be studied in more detail. The user is given reliability, and study on the reserves that allow absorbing market volatilities is imperative. The literature suggests planning with multiple agents to reduce greenhouse gas costs, separately evaluating operating expenses and revenues obtained in markets, and assessing the demand curve [10,12,13].

Another approach takes price information from a real EM to conduct energy dispatch planning [14]. However, the adequate number of IM auctions and their relationship with (PV) Gen remains undefined [11].

The main objective of this research is to evaluate the various residential solar Gen curves of an electrical MG, which is the basis for energy dispatch [15]. Gen overestimation and underestimation deviations must be adjusted using a statistical function [8]. The variations are quantified and compared with Monte Carlo simulations [16]. Once the uncertainty costs (UCs) are estimated, UCs due to deviations in the energy dispatch of the MG are evaluated [8]. The solution strategy uses the variable neighbourhood search-differential evolutionary particle swarm optimization (VNS-DEEPSO) algorithm in two stages: the first one optimizes the economic benefits of the MG, and the second one optimizes the IMs [17]. This algorithm was selected due to its high performance in smart MG optimization problems [18].

The evaluations of the revenues from solar Gen were conducted by taking 500 representative scenarios out of 5000 [14]. The costs were collated with the results obtained from the Monte Carlo simulations, yielding an error between $7 \times 10^{-5}\%$ and 0.0168% for one day of operation. The prices of uncertainty are evaluated by varying the IM auctions. It is ascertained that with greater number of auctions, the imbalances in the scheduling of solar Gen decrease.

This article presents the following structure: In Section 2, works related to the present investigation are compared. Section 3 presents the mathematical formulation of the UCs and the formulation of the objective cost function of the MG. Section 4 offers the case study, Section 5 shows the results, and Section 6 outlines the main conclusions of this investigation.

2. State of the Art

In the literature, studies on smart MGs that optimize resources are reviewed [19]. Most of the works propose the improvement of the services of the steady demand and the generators [6,19,20] so that users can participate in demand response (DR) programs [21]. They can also collaborate with flexible load management by improving their consumption habits [15,22,23]. Energy storage systems (ESSs) promise to provide further flexibility to stakeholders, who can buy off-peak energy hours and sell it during peak hours [4,10,17,23]. EMs benefit from previous integration that also facilitates the penetration of renewable resources such as solar energy [8,15]. In the case of IMs, the auction numbers play an essential role in the planning of energy dispatch [11].

This research focuses on the comparison of the storage systems (SSs) with batteries, IMs, and solar energy UCs (SEUCs), as shown in Table 1. SSs with batteries include different models of ESSs and residential EVs (REVs). The models in the table are listed below. Model 1 consists of an aggregator that performs transactions between ESSs and MPs, while Model 6 appraises the interaction between providers and users [4]. Model 2 evaluates the revenues from buying and selling in the market [6]. Model 3 suggests evaluating consumption patterns and their interaction with electricity prices [24]. Model 4 shows the incentives of IMs for intermittent generators, which can participate through meritocracy [25–31]. Model 5 estimates deviations in the energy dispatch due to EV uncertainty [8]. Model 7 encourages the participation of programs with DR; in addition, Model 8 includes IMs [19,20].

Table 1. Review of electrical MGs.

No	SS	IM	SEUC	Comments
1	Yes	No	No	MG with RS, REV, MP, DR, ESS, and Gen [4]
2	No	No	No	Include DR and Gen [6]
3	No	No	No	Include ESS and Gen [24]
4	No	Yes	No	Include DR and Gen [25]
5	No	No	Yes	Include REV and Gen [8]
6	Yes	No	No	Smart grid with DR [29]
7	No	No	No	Include DR and Gen [19]
8	No	Yes	No	Include DR and Gen [20]
9	Yes	No	No	Include ESS, demand, and Gen [26]
10	Yes	No	No	Include DR, Gen, ESS, and EV [22]
11	No	No	No	Include DR and Gen [21]
12	No	No	No	Include RS, MP, Gen, demand, and ESS [27]
13	Yes	No	No	Include RS, Gen, ESS, and REV [30]
14	No	No	No	Include DR, Gen, ESS, and REV [31]
15	No	No	No	Include DR, Gen, and ESS [12]
16	No	No	No	Include DR, Gen, and ESS [32]
17	Yes	Yes	Yes	MG with RS, REV, MP, DR, ESS, REV, BSS, and Gen(This proposal belongs to this paper)

Model 9 uses ESSs and predicts demand and Gen [26]. Model 10 enables load reduction [22]. Model 11 encourages competitiveness in EMs between DR and Gen [22]. Model 12 combines renewable sources (RSs), such as a PV panel and ESSs, to reduce CO_2 emissions [27]. Model 13 schedules the commissioning of thermal power plants, which reduces gas emissions and operating costs. Additional vehicles, wind and photovoltaic generators, and ESSs are connected to the grid. In addition, Model 14 schedules the load when faced with uncertainty regarding the future price. Model 15 reduces both costs and environmental emissions by using hybrid systems with batteries and wind and solar generators. Model 16 formulates distributed energy resources, with the energy reserve capacity and coordination of the operation with renewable resources and cogeneration. Model 17, which is the model proposed in this research, turns out to be the most complete. It has an aggregator managing the MG's resources, including RS, REV, MP, IM, DR, ESS, REV, BSS, and Gen [5,7,14,28]. In addition, the MG has restrictions that prevent the appearance of monopolies, pivotal agents, and a minimum supply of demand. [9,13]. Furthermore, the mathematical formulation of the uncertainty caused by deviations in solar energy dispatch is stated [8,15]. The costs of the MG are optimized using the VNS-DEEPSO algorithm, which presents the best performance in a similar MG [17,18].

3. Mathematical Formulation

The mathematical models are presented in two sections. The first section establishes the UCs for solar Gen. The second section formulates the objective function for the MG.

3.1. Uncertainty Costs of Photovoltaic Generation

The irradiance distribution (G) is represented using a probability function (f_G) , where the parameter (λ) is the mean, and the parameter (β) is the standard deviation [33]. The distribution for intraday solar radiation curves can be adjusted as follows.

$$f_G(G) = \frac{1}{G\beta\sqrt{2\pi}} \cdot e^{-\frac{(ln(G)-\lambda)^2}{2\beta^2}} ; 0 < G < \infty \tag{1}$$

In solar panels, the power that is generated depends on the reference irradiance R_C. The irradiance can be represented by quadratic or linear behavior, as depicted below [33].

$$f_G W_{PV}(G) = \begin{cases} W_{PV_r} \cdot \frac{G^2}{G_r R_C}, 0 < G < R_C \\ W_{PV_r} \cdot \frac{G}{G_r}, G > R_C \end{cases} \tag{2}$$

The overestimation $C_{PV,o,i}$ or underestimation $C_{PV,u,i}$ represents the deviations in the binding dispatch and IMs. The variable $W_{PV,i}$ represents PV generators i and the available power, while the power programmed by the aggregator is represented by $W_{PV,s,i}$.

$$UCF = C_{PV,u,i}(W_{PV,s,i} - W_{PV,i}) + C_{PV,o,i}(W_{PV,i} - W_{PV,s,i}) \tag{3}$$

3.2. Objective Function of Microgrid

The smart MG is represented in Figure 1, which has a bidirectional flow of information. The following tasks are undertaken: buying and selling energy in IMs, charging and discharging ESSs, charging and discharging batteries from a BSS, and charging and discharging EVs. Other elements that comprise the MG are the distributed generators (DGs) and load with DR [5,7,14,28]. In addition, the smart MG considers restrictions such as the Herfindahl–Hirschman concentration index and the index of the three most prominent bidders to avoid monopolies and pivotal agents [9]. There is also the demand welfare, which ensures a minimum consumption of the demand [13].

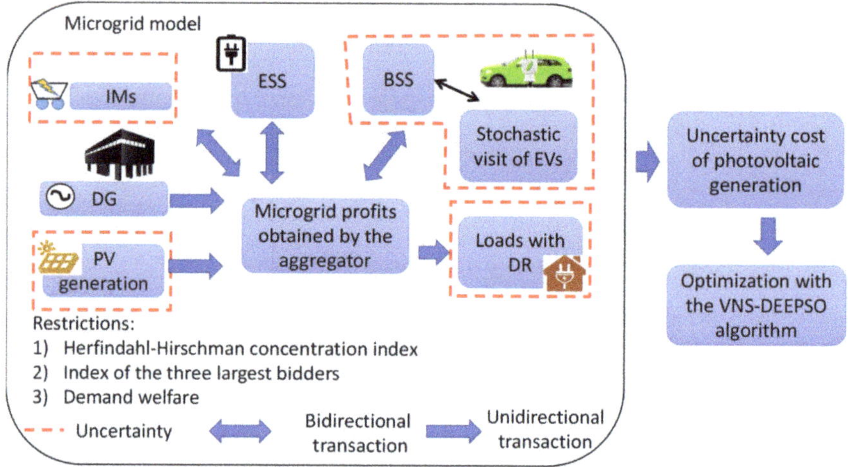

Figure 1. Structure of the electrical microgrid.

The MG model operates in a black box. Information is taken from a real MG, in which the input variables are calculated, and the MG model calculates the benefits obtained as presented below [14]. The profits of the network are represented by P, periods are represented by t, scenarios are represented by s, the probability of the occurrence of each scenario is characterized by Pr, Ns is the maximum number of scenarios, and T is the maximum number of periods.

$$MG_{Total}^{Intraday+1} = \sum_{s=1}^{N_s} \left(\sum_{t=1}^{T=Ti} P_{(t,s)} + \sum_{t=Ti+1}^{T=2Ti} P_{(t,s)} + \sum_{t=iTi+1}^{T=24} P_{(t,s)} \right) \cdot \Pr(s) \tag{4}$$
$$\{Ti, 2Ti, \ldots, 24\} \epsilon \mathbb{Z}$$

The deviations in the dispatch of solar energy will appear in each of the intraday periods. The UCs are calculated for each period, where N_{PV} represents the maximum number of PV Gen units.

$$UC_{Total}^{Intraday+1} = \begin{array}{l} \sum\limits_{s=1}^{N_s}\left(\sum\limits_{t=1}^{T=Ti}\left(\sum\limits_{j=1}^{N_{PV}} C_{PV,u,\,ju}(W_{PV,s,i} - W_{PV,i})\right) + \ldots\right. \\ \ldots \sum\limits_{t=Ti+1}^{T=2Ti}\left(\sum\limits_{j=1}^{N_{PV}} C_{PV,u,\,ju}(W_{PV,s,i} - W_{PV,i})\right) + \ldots \\ \ldots + \sum\limits_{t=iTi+1}^{T=24}\left(\sum\limits_{j=1}^{N_{PV}} C_{PV,u,\,ju}(W_{PV,s,i} - W_{PV,i})\right) + \ldots \\ \ldots + \sum\limits_{t=1}^{T=Ti}\left(\sum\limits_{j=1}^{N_{PV}} C_{PV,o,\,ju}(W_{PV,i} - W_{PV,s,i})\right) + \ldots \\ \ldots \sum\limits_{t=Ti+1}^{T=2Ti}\left(\sum\limits_{j=1}^{N_{PV}} C_{PV,o,\,ju}(W_{PV,i} - W_{PV,s,i})\right) + \ldots \\ \left.\ldots + \sum\limits_{t=iTi+1}^{T=24}\left(\sum\limits_{j=1}^{N_{PV}} C_{PV,o,\,ju}(W_{PV,i} - W_{PV,s,i})\right)\right) \\ \{Ti, 2Ti, \ldots, 24\}\epsilon\mathbb{Z} \end{array} \tag{5}$$

The objective function is defined as minimizing the costs of uncertainty for the dispatch of solar energy minus the benefits obtained in the MG, which is optimized by using the VNS-DEEPSO algorithm.

$$minimize\ Z_{Total}^{Intraday+1} = UC_{Total}^{Intraday+1} - MG_{Total}^{Intraday+1} \tag{6}$$

4. Case Study Presentation

The case study is presented in two sections. The first section shows the statistical data used to evaluate the costs of uncertainty, and in the second one, the MG is given.

4.1. Residential Solar Generators

The power that is generated daily is taken from [14], where the energy for residential solar panels is considered. Figure 2 shows the power supply for 500 representative scenarios.

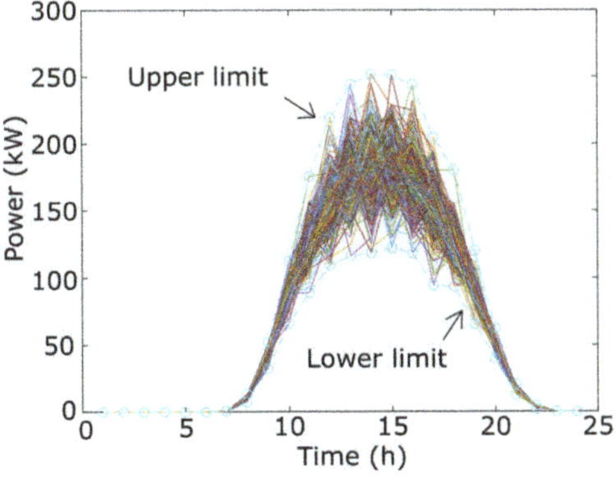

Figure 2. Residential PV generators in 24 h, data from [14].

The UC function must be validated using Monte Carlo simulations with which random irradiance values are obtained by assuming the proportionality between the irradiance and the generated power. Solar radiation parameters are considered according to solar radiation distribution frequency functions, as shown in Table 2 [34]. The penalties for overestimation and underestimation are considered [8].

Table 2. Solar generation parameter, data from [34].

Symbol	Parameter	Value
W_{PVr}	PV Gen source (MW)	100
G_r	Nominal irradiance (W/m^2)	775
R_C	Irradiance of reference (W/m^2)	116
$W_{PV,\infty}$	Maximum power Gen (MW)	150
N	Number of iterations	100,000
$W_{PV,s,i}$	Scheduled PV Gen (MW)	100
$C_{PV,u,i}$	Penalty for underestimation (USD/MW)	300
$C_{PV,o,i}$	Penalty for overestimation (USD/MW)	700

The solar radiation values in Figure 2 are the basis for calculating the mean and standard deviation of each hour. The obtained values are summarized in Tables 3 and 4.

Table 3. Mean and standard deviation between 8 and 14 h.

Symbol	8	9	10	11	12	13	14
β	0.0965	0.0710	0.0673	0.0887	0.1030	0.1069	0.1121
λ	2.3059	3.7685	4.4929	4.8512	5.0494	5.1577	5.2085

Table 4. Mean and standard deviation between 15 and 21 h.

Symbol	15	16	17	18	19	20	21
β	0.0965	0.0710	0.0673	0.0887	0.1030	0.1069	0.1121
λ	2.3059	3.7685	4.4929	4.8512	5.0494	5.1577	5.2085

4.2. *Objective Function of Microgrid*

MG is located in Portugal and comprises 17 solar Gen units, 5 dispatchable units, 34 REVs, 2 ESSs, an external electricity service provider, and 90 users who actively participate in DR programs [35,36]. The distribution transformer is 160 kVA and connects to a medium and low voltage line of 30 kV/400 V–230 V [37]. The five dispatchable units comprise four DGs and an external solar generator. The transformer is connected to 25 buses [37]. Additionally, MG can transfer energy with intraday markets [11]. It also has penalties for costs of uncertainty in the Gen of solar energy, market power restrictions, and constraints on the minimum supply of energy for demand [9,13,15]. MG is optimized using the VNS-DEEPSO algorithm, which improves MG in the first stage and MIs in the second stage, as shown in Figure 3 [17,18].

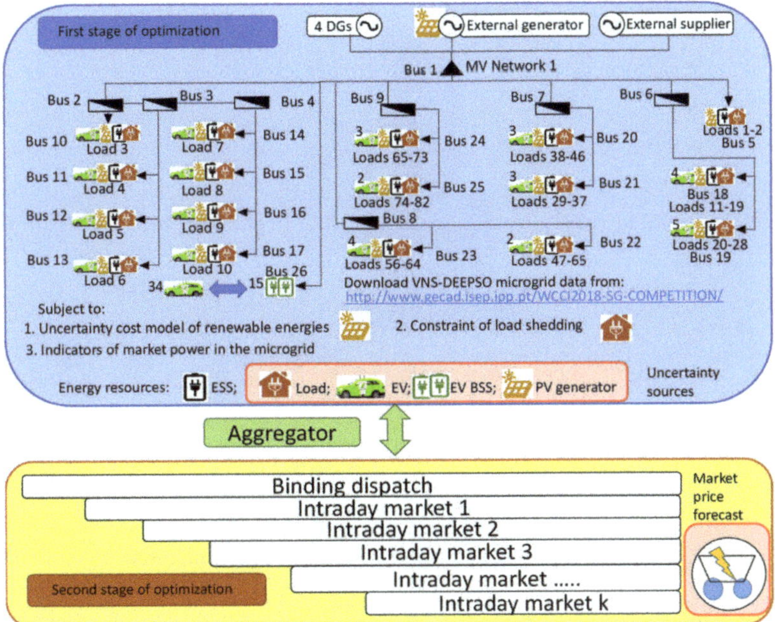

Figure 3. Solar generators in the electrical microgrid [14].

5. Results and Discussion

The results are presented in two sections. In the first section, the UCs for residential solar energy Gen are validated, while in the second one, the UCs are estimated by varying the IMs of the MG.

5.1. Uncertainty Costs with Residential Solar Generators

The validation uses Monte Carlo simulations to determine the histograms of irradiance and solar power generated with the underestimated and overestimated costs of solar radiation, as shown in Figure 4.

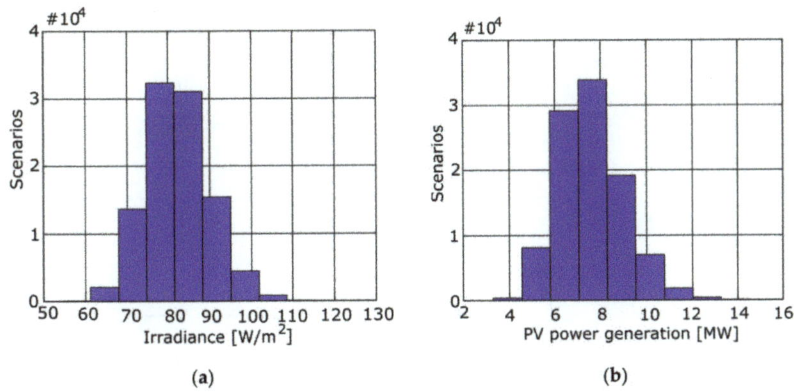

Figure 4. Histograms of (**a**) irradiance and (**b**) solar power generated. (# means 10^4).

Penalties due to UCs are determined, and UCs are calculated while solar Gen varies (Figure 5).

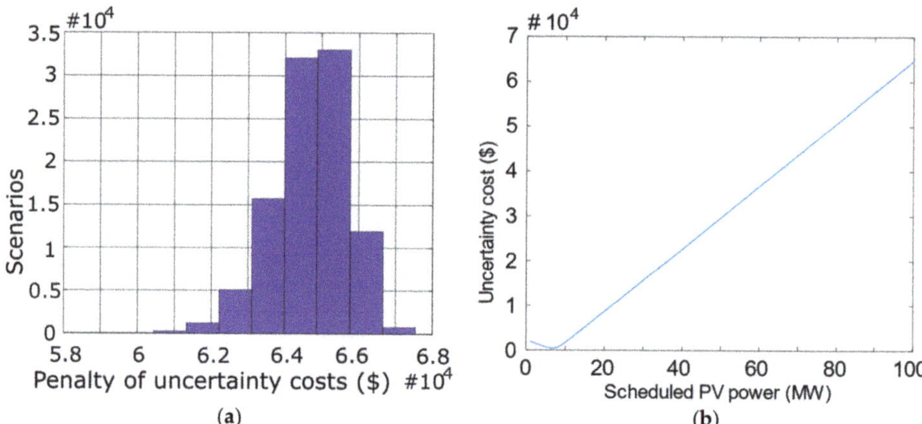

Figure 5. Monte Carlo simulations for (**a**) UC function histogram and (**b**) evaluation of UCs by varying solar power. (# means 10^4).

Finally, the UCs are evaluated for each hour using Monte Carlo simulations and the UC function. In the Monte Carlo simulations, the value is obtained with the average of the estimated values MC_{PV} [38]. The costs for underestimation and overestimation are used [33] as shown below.

$$AC_{PV} = E[C_{PV,u,i}(W_{PV,s,i}, W_{PV,i})] + E[C_{PV,u,i}(W_{PV,o,i}, W_{PV,i})] \tag{7}$$

The estimated error when evaluating the Monte Carlo functions and the UC function is summarized in Tables 5 and 6. The error is in the range between 7×10^{-5}% and 0.0168%. This research differs from previous works in which the uncertainty costs of renewable energies per day had been evaluated; in this research, a set of intraday evaluations per hour is carried out. For comparative purposes, the highest error reported in each research is taken, in which the error of 0.0168% is more exact than the errors obtained for 0.0615% from [39], 0.0343% from [16], and 0.072% from [33]. This means that this investigation contains the error closest to zero.

Table 5. The estimated error between 8 and 14 h.

Symbol	8	9	10	11	12	13	14
MC_{PV} ($)	64,698	68,525	63,725	58,465	55,843	54,220	53,375
AC_{PV} ($)	64,695	68,524	63,723	58,460	55,842	54,215	53,383
e ($)	0.0049	4×10^{-4}	0.0026	0.0099	0.0027	0.0088	0.0151

Table 6. The estimated error between 15 and 21 h.

Symbol	15	16	17	18	19	20	21
MC_{PV} ($)	69,920	68,526	63,726	58,465	55,844	54,216	53,374
AC_{PV} ($)	69,920	68,524	63,723	58,460	55,842	54,215	53,383
e ($)	7×10^{-5}	0.002	0.0047	0.0091	0.0041	0.0022	0.0168

5.2. Uncertainty Costs Varing Intraday Markets in the Microgrid

The uncertainty from the solar energy dispatch is due to the underestimation and overestimation of the power. The MG model considers the UCs for solar Gen with 2, 3, 4, and 6 IMs. The auctions are taken in symmetrical times; for example, in the case of three intraday markets, the auctions are conducted every 8 h, as shown in [11]. The uncertainty costs due to overrating and underestimating the solar energy dispatch are reduced with a more significant number of intraday markets, as shown in Figure 6.

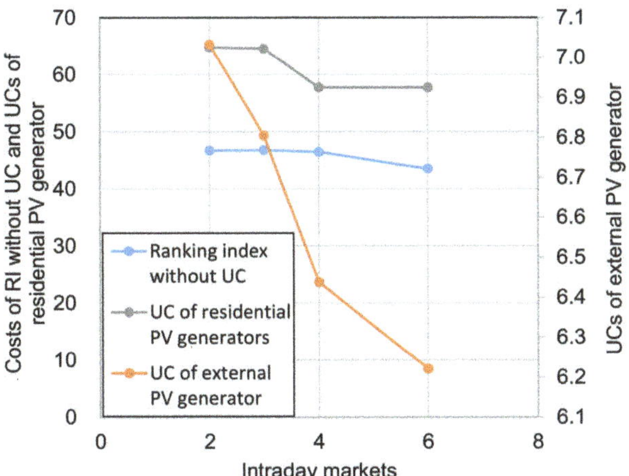

Figure 6. UCs of PV generators varying IMs.

6. Conclusions

The uncertainty in energy distribution planning is inevitable and can have minor impacts on the planning of electrical microgrids. This research quantified the uncertainty of the solar generation of an electric microgrid and validated the methodology using Monte Carlo simulations. The relative error of the uncertainty cost estimation function for the solar energy generation obtained values in the range between $7 \times 10^{-5}\%$ and 0.0168%. In addition, the authors evaluated the effect of intraday markets in an optimization case of an electrical microgrid. It was found that the costs of uncertainty for the generation of the solar energy decrease when the number of intraday markets increases, thus considering symmetrical daily auction periods. For example, in the case of three intraday markets, the auctions are carried out every 8 h. The aggregator improved the economic management of the network with a more significant number of intraday markets. Future works should evaluate other sources of clean energy such as micro-hydroelectric plants and wind turbines. Such studies must also assess the effect of the implementation of intraday markets with non-symmetrical daily auction periods on the costs of uncertainty.

Author Contributions: This research was carried out with the following contributions by the authors: conceptualization, D.A. and S.R.; methodology, J.G.-G. supported by Colciencias, Colombia, Call 785 for national doctorates; validation, S.R. and J.G.-G.; formal analysis, D.A.; investigation, J.G.-G.; supervision, D.A. and S.R.; writing—original draft preparation, J.G.-G.; writing—review and editing, D.A. and S.R.; funding acquisition, S.R. All authors have read and agreed to the published version of the manuscript.

Funding: This research received no external funding.

Data Availability Statement: The MG model is available online at: http://www.gecad.isep.ipp.pt/WCCI2018-SG-COMPETITION/ by the group GECAD—Polytechnic of Porto (access date 16/03/2021).

Conflicts of Interest: The authors declare no conflict of interest.

References

1. Gharavi, H.; Ghafurian, R. Smart grid: The electric energy system of the future. *Proc. IEEE* **2011**, *9*, 913–914.
2. Marzband, M.; Sumper, A.; Domínguez-García, J.L.; Gumara-Ferret, R. Experimental validation of a real time energy management system for microgrids in islanded mode using a local day-ahead electricity market and MINLP. *Energy Convers. Manag.* **2013**, *76*, 314–322. [CrossRef]

3. Garcia, J.; Alvarez, D.; Rivera, S. Ensemble Based Optimization for Electric Demand Forecast: Genetic Programming and Heuristic Algorithms. *Rev. Int. Métodos Numér. Cálc. Diseño Ing.* **2020**, *1*, 1–13. [CrossRef]
4. Soares, J.; Fotouhi Ghazvini, M.A.; Borges, N.; Vale, Z. A stochastic model for energy resources management considering demand response in smart grids. *Electr. Power Syst. Res.* **2017**, *143*, 599–610. [CrossRef]
5. Garcia-Guarin, J.; Infante, W.; Alvarez, D.; Rivera, S. Scheduling optimization for smart microgrids considering twolevels transactions of electric vehicles and energy markets. *J. Phys. Conf. Ser.* **2020**, *1708*, 012019. [CrossRef]
6. Abrishambaf, O.; Faria, P.; Spínola, J.; Vale, Z. An Aggregation Model for Energy Resources Management and Market Negotiations. *Adv. Sci. Technol. Eng. Syst. J.* **2018**, *3*, 231–237. [CrossRef]
7. Garcia-Guarin, J.; Infante, W.; Ma, J.; Alvarez, D.; Rivera, S. Optimal Scheduling of Smart Microgrids Considering Electric Vehicle Battery Swapping Stations. *Int. J. Electr. Comput. Eng.* **2020**, *10*, 5093–5107.
8. Arévalo, J.; Santos, F.; Rivera, S. Application of Analytical Uncertainty Costs of Solar, Wind and Electric Vehicles in Optimal Power Dispatch. *Ingeniería* **2017**, *22*, 324–346. [CrossRef]
9. Garcia-Guarin, J.; Rivera, S.; Trigos, L. Multiobjective optimization of smart grids considering market power. *J. Phys. Conf. Ser.* **2019**, *1409*, 012006. [CrossRef]
10. Lezama, F.; Soares, J.; Faia, R.; Pinto, T.; Vale, Z. A New Hybrid-Adaptive Differential Evolution for a Smart Grid Application under Uncertainty. In Proceedings of the 2018 IEEE Congress on Evolutionary Computation, CEC 2018—Proceedings, Rio de Janeiro, Brazil, 8–13 July 2018; pp. 1–8.
11. Garcia-Guarin, J.; Duran-Pinzón, M.; Paez-Arango, J.; Rivera, S. Energy planning for aquaponics production considering intraday markets. *Arch. Electr. Eng.* **2020**, *69*, 89–100. [CrossRef]
12. Radhakrishnan, B.M.; Srinivasan, D.; Mehta, R. Fuzzy-Based Multi-Agent System for Distributed Energy Management in Smart Grids. *Int. J. Uncertain. Fuzziness Knowl.-Based Syst.* **2016**, *24*, 781–803. [CrossRef]
13. Garcia-Guarin, J.; Alvarez, D.; Bretas, A.; Rivera, S. Schedule Optimization in A Smart Microgrid Considering Demand Response Constraints. *Energies* **2020**, *13*, 4567. [CrossRef]
14. Lezama, F.; Soares, J.; Vale, Z.; Rueda, J.; Wagner, M. CEC/GECCO 2019 Competition Evolutionary Computation in Uncertain Environments: A Smart Grid Application. In Proceedings of the 2018 IEEE Congress on Evolutionary Computation, CEC 2018—Proceedings, Rio de Janeiro, Brazil, 8–13 July 2018.
15. Garcia Guarin, J.; Lezama, F.; Soares, J.; Rivera, S. Operation scheduling of smart grids considering stochastic uncertainty modelling. *Far East J. Math. Sci.* **2019**, *1*, 77–98. [CrossRef]
16. Vargas, S.; Rodriguez, D.; Rivera, S. Mathematical Formulation and Numerical Validation of Uncertainty Costs for Controllable Loads. *Rev. Int. Métodos Numéricos Para Cálculo Y Diseño En Ing.* **2019**, *35*, 1–17. [CrossRef]
17. García-Guarín, P.J.; Cantor-López, J.; Cortés-Guerrero, C.; Guzmán-Pardo, M.A.; Rivera, S. Implementación del algoritmo VNS-DEEPSO para el despacho de energía en redes distribuidas inteligentes. *INGE CUC* **2019**, *15*, 142–154. [CrossRef]
18. Garcia-Guarin, J.; Rodriguez, D.; Alvarez, D.; Rivera, S.; Cortes, C.; Guzman, A.; Bretas, A.; Aguero, J.R.; Bretas, N. Smart microgrids operation considering a variable neighborhood search: The differential evolutionary particle swarm optimization algorithm. *Energies* **2019**, *12*, 3149. [CrossRef]
19. Jadid, S.; Zakariazadeh, A. Energy and reserve scheduling of microgrid using multi-objective optimization. In Proceedings of the 22nd International Conference and Exhibition on Electricity Distribution (CIRED 2013), Stockholm, Sweden, 10–13 June 2013; pp. 631–660.
20. Cecati, C.; Citro, C.; Siano, P. Combined Operations of Renewable Energy Systems and Responsive Demand in a Smart Grid. *IEEE Trans. Sustain. Energy* **2011**, *2*, 468–476. [CrossRef]
21. Faria, P.; Soares, T.; Vale, Z.; Morais, H. Distributed generation and demand response dispatch for a virtual power player energy and reserve provision. *Renew. Energy* **2014**, *66*, 686–695. [CrossRef]
22. Sousa, T.; Morais, H.; Vale, Z.; Faria, P.; Soares, J. Intelligent Energy Resource Management Considering Vehicle-to-Grid: A Simulated Annealing Approach. *IEEE Trans. Smart Grid* **2012**, *3*, 535–542. [CrossRef]
23. Garcia-Guarin, J.; Rivera, S.; Rodriguez, H. Smart grid review: Reality in Colombia and expectations. *J. Phys. Conf. Ser.* **2019**, *1257*, 012011. [CrossRef]
24. Tushar, W.; Chai, B.; Yuen, C.; Smith, D.B.; Wood, K.L.; Yang, Z.; Poor, H.V. Three-Party Energy Management with Distributed Energy Resources in Smart Grid. *IEEE Trans. Ind. Electron.* **2015**, *62*, 2487–2498. [CrossRef]
25. Zhou, Y.; Wang, C.; Wu, J.; Wang, J.; Cheng, M.; Li, G. Optimal scheduling of aggregated thermostatically controlled loads with renewable generation in the intraday electricity market. *Appl. Energy* **2017**, *188*, 456–465. [CrossRef]
26. Matthiss, B.; Momenifarahaniy, A.; Ohnmeissz, K.; Felderx, M. Influence of Demand and Generation Uncertainty on the Operational Efficiency of Smart Grids. In Proceedings of the 7th International IEEE Conference on Renewable Energy Research and Applications, ICRERA 2018, Paris, France, 14–17 October 2018; pp. 751–756.
27. Di Somma, M.; Graditi, G.; Heydarian-Forushani, E.; Shafie-khah, M.; Siano, P. Stochastic optimal scheduling of distributed energy resources with renewables considering economic and environmental aspects. *Renew. Energy* **2018**, *116*, 272–287. [CrossRef]
28. Lezama, F.; Soares, J.; Vale, Z.; Rueda, J.; Rivera, S.; Elrich, I. 2017 IEEE competition on modern heuristic optimizers for smart grid operation: Testbeds and results. *Swarm Evol. Comput.* **2019**, *44*, 420–427. [CrossRef]
29. Chen, J.; Yang, B.; Guan, X. Optimal demand response scheduling with Stackelberg game approach under load uncertainty for smart grid. *IEEE SmartGridComm* **2012**, *1*, 546–551. [CrossRef]

30. Saber, A.Y.; Venayagamoorthy, G.K. Resource scheduling under uncertainty in a smart grid with renewables and plug-in vehicles. *IEEE Syst. J.* **2012**, *6*, 103–109. [CrossRef]
31. Deng, R.; Yang, Z.; Chen, J.; Chow, M.-Y. Load Scheduling With Price Uncertainty and Temporally-Coupled Constraints in Smart Grids. *IEEE Trans. Power Syst.* **2014**, *29*, 2823–2834. [CrossRef]
32. Quashie, M.; Marnay, C.; Bouffard, F.; Jóos, G. Optimal planning of microgrid power and operating reserve capacity. *Appl. Energy* **2018**, *210*, 1229–1236. [CrossRef]
33. Arevalo, J.C.; Santos, F.; Rivera, S. Uncertainty cost functions for solar photovoltaic generation, wind energy generation, and plug-in electric vehicles: Mathematical expected value and verification by Monte Carlo simulation. *Int. J. Power Energy Convers.* **2019**, *10*, 171–207. [CrossRef]
34. Chang, T.P. Investigation on frequency distribution of global radiation using different probability density functions. *Int. J. Appl. Sci. Eng.* **2010**, *8*, 99–107.
35. Lezama, F.; Sucar, L.E.; de Cote, E.M.; Soares, J.; Vale, Z. Differential evolution strategies for large-scale energy resource management in smart grids. In Proceedings of the Genetic and Evolutionary Computation Conference Companion—GECCO '17, Berlin, Germany, 15–19 July 2017; pp. 1279–1286. [CrossRef]
36. Soares, J.; Silva, M.; Canizes, B.; Vale, Z. MicroGrid der control including EVs in a residential area. In *Proceedings of the 2015 IEEE Eindhoven PowerTech, PowerTech 2015*; Institute of Electrical and Electronics Engineers Inc.: Piscataway, NJ, USA, 2015; pp. 1–6.
37. Canizes, B.; Silva, M.; Faria, P.; Ramos, S.; Vale, Z. Resource scheduling in residential microgrids considering energy selling to external players. In Proceedings of the 2015 Clemson University Power Systems Conference, PSC 2015, Clemson, SC, USA, 10–13 March 2015.
38. Wackerly, D.; Mendenhall, W.; Scheaffer, R. *Estadística Matemática con Aplicaciones*, 7th ed.; Cengage Learning: Santa Fe, Mexico, 2010; ISBN 970-625-016-6.
39. Molina Sanchez, F.S.; Pérez Sichacá, S.J.; Rivera Rodriguez, S.R. Formulación de Funciones de Costo de Incertidumbre en Pequeñas Centrales Hidroeléctricas dentro de una Microgrid. *Ing. USBMed* **2017**, *8*, 29–36. [CrossRef]

MDPI

Article

Exploiting the S-Iteration Process for Solving Power Flow Problems: Novel Algorithms and Comprehensive Analysis

Marcos Tostado-Véliz [1,*], Salah Kamel [2], Ibrahim B. M. Taha [3] and Francisco Jurado [1]

1 Department of Electrical Engineering, University of Jaén, 23700 Linares, Spain; fjurado@ujaen.es
2 Department of Electrical Engineering, Faculty of Engineering, Aswan University, Aswan 81542, Egypt; skamel@aswu.edu.eg
3 Department of Electrical Engineering, College of Engineering, Taif University, Taif 21944, Saudi Arabia; i.taha@tu.edu.sa
* Correspondence: mtostado@ujaen.es

Abstract: In recent studies, the competitiveness of the Newton-S-Iteration-Process (Newton-SIP) techniques to efficiently solve the Power Flow (PF) problems in both well and ill-conditioned systems has been highlighted, concluding that these methods may be suitable for industrial applications. This paper aims to tackle some of the open topics brought for this kind of techniques. Different PF techniques are proposed based on the most recently developed Newton-SIP methods. In addition, convergence analysis and a comparative study of four different Newton-SIP methods PF techniques are presented. To check the features of considered PF techniques, several numerical experiments are carried out. Results show that the considered Newton-SIP techniques can achieve up to an eighth order of convergence and typically are more efficient and robust than the Newton–Raphson (NR) technique. Finally, it is shown that the overall performance of the considered PF techniques is strongly influenced by the values of parameters involved in the iterative procedure.

Keywords: power flow; S-iteration process; Newton–Raphson; high order newton-like method; computational efficiency

Citation: Tostado-Véliz, M.; Kamel, S.; Taha, I.B.M.; Jurado, F. Exploiting the S-Iteration Process for Solving Power Flow Problems: Novel Algorithms and Comprehensive Analysis. *Electronics* **2021**, *10*, 3011. https://doi.org/10.3390/electronics10233011

Academic Editors: Marinko Barukčić, Nebojša Raičević and Vasilija Šarac

Received: 2 November 2021
Accepted: 29 November 2021
Published: 2 December 2021

Publisher's Note: MDPI stays neutral with regard to jurisdictional claims in published maps and institutional affiliations.

1. Introduction

Power Flow (PF) is the backbone of power system analysis. From a mathematical point of view, PF is a nonlinear problem in which the operational steady state of a power system is obtained. Traditional methods for tackling this problem are the iterative NR [1] and decoupled techniques [2–4].

Although PF is customarily solved in polar coordinates form, other formulations have been studied. A PF formulation based on current injections instead of power injections has been proposed by da Costa et al. [5] and posteriorly embellished by Garcia et al. in [6]. Saleh has developed a formulation of the PF problem in the well-known d-q framework in [7,8]. More recently, PF formulation in complex variables has been exploited in [9], using Witinger Calculus.

Ill-conditioned systems bring some issues for traditional PF solution techniques. This topic has been profusely studied for decades. For example, the reader can be referred to the works of Iwamoto and Tamura [10], Tripathy et al. [11] or Braz et al. [12]. More recently, these kinds of problems have been tackled using the Continuous Newton's paradigm by Milano in [13] or by some of the authors in several recent papers [14–16]. The works of Pourbagher and Derakhshandeh have been focused on the solution of ill-conditioned power systems using the Levenberg–Marquardt technique [17,18]. Alternatively, a novel paradigm has been proposed by the authors in [19], which studies the application of the Gauss–Newton method for PF analysis.

High-order Newton-like methods have also been studied for PF analysis. In [20], Pourbagher and Derakhshandeh studied the application of Newton-like techniques of

3rd, 4th, and 5th order to PF analysis. On the other hand, a Newton-like technique with a superquadratic convergence rate has been proposed to solve the PF problem in well-conditioned systems by some of the authors in [21].

Regarding continuation, Homotopic and Holomorphic techniques have also been exploited for PF analysis. The well-known Continuation Power Flow [22] may be the greatest exponent of this kind of methodology. This approach is traditionally used to determine the stability margin of a power system by calculating its Maximum Loadability Point. Some recent efforts have been made for adapting Continuation Power Flow to distribution systems [23,24]. The Homotopic principle has been applied to PF analysis by Yang and Zhou in [25]. Posteriorly, a family of robust and efficient PF solution techniques based on a combined Newton–Homotopic approach has been developed by some of the authors in [26]. The PF solution by the Holomorphic Embedding method was firstly studied in [27]. Recently, the PF solution by this principle has been further studied in [28,29].

The application of the S-iteration process (SIP) [30] to PF analysis has been recently tackled by the authors in [31]. In this regard, an iterative algorithm based on a combined Newton-SIP approach developed in [32] was adapted for solving either well or ill-conditioned systems. The developed solver turned out to be very efficient, since only an LU decomposition is required in the whole iterative process. This was reflected in very promising results, frequently outperforming NR or the decoupled methods. In addition, it turned out to be quite robust, efficiently handling some large and very large ill-conditioned systems. However, due to the linear convergence characteristic of this method, it suffers from slow convergence in heavy loading cases. In order to overcome this drawback, a Jacobian updated mechanism has also been proposed. Definitely, the PF solution technique proposed in [31] and its variant can be widely used in industry tools due to its capacity for managing well and ill-conditioned equations and its simplicity and efficiency. However, the application of SIP for PF analysis is still far from being fully studied. For example, several topics still need to be further analyzed:

- The reference [31] is limited to studying only one of the algorithms developed in [32]. In the latter reference, along with [33], three other Newton-SIP methods were developed. The applicability of these techniques to PF analysis has not been studied yet.
- Although the Jacobian updated mechanism proposed in [31] allows overcoming the slow-convergence issues in heavy loading systems, the whole iterative procedure remains linear. Consequently, many iterations are normally employed to achieve a feasible solution.
- The overall performance of the Newton-SIP technique studied in [31] strongly depends on the value of the parameters involved in the iterative procedure (s-parameters).

In order to respond to the issues above, the authors strongly believe that further analysis of the SIP applied to PF analysis is still required. This paper aims to fill this gap by profusely studying the Newton-like methods developed in [32,33]. Two schemes are considered. Firstly, we take the constant Jacobian matrix, which corresponds with the standard form of the techniques developed in [32,33]. This mechanism brings linear algorithms; hence, the Newton-SIP methods are also studied for a fully updated scheme in which the Jacobian matrix involved is updated each iteration. The developed methods are compared in terms of efficiency and convergence rate. Finally, we study several numerical experiments in order to analyze the performance of the different Newton-SIP methods in well and ill-conditioned systems, comparing their results with those obtained by NR and analyzing the influence of the s-parameters in the overall performance of the Newton-SIP approaches.

The remainder of the paper is organized as follows. Firstly, the Newton-SIP methods developed in [32,33] are presented and adapted to the PF problem in Section 2. A convergence study of the considered PF solution techniques is provided in Section 3. Section 4 compares the studied methodologies in terms of efficiency. Section 5 describes the different numerical experiments considered, and the results obtained are interpreted and discussed. Finally, Section 6 concludes the paper.

2. Newton-SIP Methods Applied to PF Analysis

2.1. Background

The PF problem in polar coordinates can be established as a set of n nonlinear equations given by [34]:

$$\mathbf{g}(\boldsymbol{x}) = \begin{cases} P_i^{sch} - \sum |V_i||V_j||Y_{ij}|\cos(\theta_{ij} - \delta_i + \delta_j) = 0 \\ Q_i^{sch} - \sum |V_i||V_j||Y_{ij}|\sin(\theta_{ij} - \delta_i + \delta_j) = 0 \end{cases} \tag{1}$$

where P_i^{sch} and Q_i^{sch} are the scheduled active and reactive power at i^{th} bus, respectively; $V_i \angle \delta_i$ is the complex voltage at the ith bus; $Y_{ij} \angle \theta_{ij}$ is the ij^{th} element of the admittance matrix; and $\boldsymbol{x} \in \mathbb{R}^n$ is the PF vector of unknowns, which is defined in polar coordinates as follows:

$$\boldsymbol{x} = \left[\boldsymbol{\delta_{PV}} | \boldsymbol{\delta_{PQ}} | \boldsymbol{V_{PQ}} \right]^T \tag{2}$$

where $\boldsymbol{\delta_{PV}} \in \mathbb{R}^{n_g}$ is the vector of voltage angles at PV buses, $\boldsymbol{\delta_{PQ}} \in \mathbb{R}^{n_l}$ is the vector of voltage angles at PQ buses, $\boldsymbol{V_{PQ}} \in \mathbb{R}^{n_l}$ is the vector of voltage magnitudes at PQ buses, and n_l and n_g are the total number of PQ and PV buses, respectively.

In the formulation above, only the well-known constant power loads have been considered, nevertheless, other type of consumers could be considered following the formulation described in the Appendix A. Explicit solutions of the system of nonlinear equations (1) cannot be directly obtained. In this regard, iterative methods are undoubtedly the most popular techniques for solving this kind of problem. Among them, the NR method has been the most widely used in PF analysis. The generic k^{th} iteration of the NR for solving (1) is given by:

$$\boldsymbol{x}^{(k+1)} = \boldsymbol{x}^{(k)} - \left[\boldsymbol{J}\left(\boldsymbol{x}^{(k)} \right) \right]^{-1} \mathbf{g}\left(\boldsymbol{x}^{(k)} \right) \tag{3}$$

where $\boldsymbol{J} \in \mathbb{R}^{n \times n}$ is the Jacobian matrix, which is formed by the first partial derivatives of (1) with respect to (2). It is well known that the NR method has local quadratic convergence.

Alternatively, other robust and high-order Newton-like methods have been proposed to solve the PF and overcome the drawbacks posed by NR (See Section 1). This paper is focused on the family of Newton-SIP methods, which are described in the following subsections.

2.2. Newton-SIP Methods (Type 1)

In [32], two Newton-SIP methods have been developed. Firstly, we denote SIP1-J$_0$ to the methodology whose generic k^{th} iteration for solving the PF is carried out as follows:

$$\begin{cases} \boldsymbol{y}^{(k)} = (1-\alpha)\boldsymbol{x}^{(k)} + \alpha\left(\boldsymbol{x}^{(k)} - \left[\boldsymbol{J}\left(\boldsymbol{x}^{(0)} \right) \right]^{-1} \mathbf{g}\left(\boldsymbol{x}^{(k)} \right) \right) \\ \boldsymbol{x}^{(k+1)} = \boldsymbol{y}^{(k)} - \left[\boldsymbol{J}\left(\boldsymbol{x}^{(0)} \right) \right]^{-1} \mathbf{g}\left(\boldsymbol{y}^{(k)} \right) \end{cases} \tag{4}$$

where α. On the other hand, another Newton-SIP method, namely SIP2-J$_0$ has been proposed in [32]. In this case, the generic k^{th} iteration of SIP2-J$_0$ for solving the PF is given by:

$$\begin{cases} \boldsymbol{y}^{(k)} = (1-\alpha)\boldsymbol{x}^{(k)} + \alpha\left(\boldsymbol{x}^{(k)} - \left[\boldsymbol{J}\left(\boldsymbol{x}^{(0)} \right) \right]^{-1} \mathbf{g}\left(\boldsymbol{x}^{(k)} \right) \right) \\ \boldsymbol{x}^{(k+1)} = \boldsymbol{y}^{(k)} - \left[\boldsymbol{J}\left(\boldsymbol{y}^{(0)} \right) \right]^{-1} \mathbf{g}\left(\boldsymbol{y}^{(k)} \right) \end{cases}. \tag{5}$$

Both SIP1-J$_0$ and SIP2-J$_0$ are defined by only one s-parameter (namely α). The main difference between SIP1-J$_0$ and SIP2-J$_0$ lies in the latter requiring two Jacobian evaluations. It is noteworthy that the iterative algorithms defined by (4) and (5) only evaluate the Jacobian at $\boldsymbol{x}^{(0)}$ and $\boldsymbol{y}^{(0)}$; hence, they are a priori more efficient than NR.

2.3. Newton-SIP Methods (Type 2)

In [33], two other Newton-SIP algorithms are proposed. Firstly, let us denote SIP3-J$_0$ to that method whose generic k^{th} iteration for solving the PF is carried out as follows:

$$\begin{cases} z^{(k)} = (1-\theta)x^{(k)} + \theta\left(x^{(k)} - \left[J\left(x^{(0)}\right)\right]^{-1} g\left(x^{(k)}\right)\right) \\ y^{(k)} = (1-\alpha)\left(x^{(k)} - \left[J\left(x^{(0)}\right)\right]^{-1} g\left(x^{(k)}\right)\right) + \alpha\left(z^{(k)} - \left[J\left(x^{(0)}\right)\right]^{-1} g\left(z^{(k)}\right)\right) \\ x^{(k+1)} = y^{(k)} - \left[J\left(x^{(0)}\right)\right]^{-1} g\left(y^{(k)}\right) \end{cases} \quad (6)$$

where θ. Finally, the methodology denoted SIP4-J_0 proposed in [33] is carried out at its generic k^{th} iteration for solving the PF problem as follows:

$$\begin{cases} z^{(k)} = (1-\theta)x^{(k)} + \theta\left(x^{(k)} - \left[J\left(x^{(0)}\right)\right]^{-1} g\left(x^{(k)}\right)\right) \\ y^{(k)} = (1-\alpha)\left(x^{(k)} - \left[J\left(x^{(0)}\right)\right]^{-1} g\left(x^{(k)}\right)\right) + \alpha\left(z^{(k)} - \left[J\left(z^{(0)}\right)\right]^{-1} g\left(z^{(k)}\right)\right) \\ x^{(k+1)} = y^{(k)} - \left[J\left(y^{(0)}\right)\right]^{-1} g\left(y^{(k)}\right) \end{cases} . \quad (7)$$

As in the previous subsection, the main difference between SIP3-J_0 and SIP4-J_0 lies in the total number of Jacobian evaluations. While the latter requires three Jacobian evaluations, SIP3-J_0 only requires one Jacobian evaluation. An important difference between the methodologies proposed in [32] and those developed in [33] is the number of s-parameters involved. While SIP1-J_0 and SIP2-J_0 are defined by only one s-parameter (α), SIP3-J_0 and SIP4-J_0 are characterized by a pair of s-parameters (α, θ). Finally, all studied Newton-SIP methods only evaluate the Jacobian matrix at the first iteration (in just one or various points); in this paper, we have considered alternative procedures in which the Jacobian matrices are updated each iteration (as in the standard NR). These alternative techniques have been called SIP1-J, SIP2-J, SIP3-J, and SIP4-J for the SIP1-J_0, SIP2-J_0, SIP3-J_0, and SIP4-J_0, respectively. With the aim to summarize, Table 1 collects the main characteristics of the studied PF solution techniques.

Table 1. Main features of the studied PF solution methods.

Method	Jacobian Evaluations	Function Evaluations	S-Parameters
NR	K	K	–
SIP1-J_0	1	$2 \times K$	α
SIP1-J	K	$2 \times K$	α
SIP2-J_0	2	$2 \times K$	α
SIP2-J	$2 \times K$	$2 \times K$	α
SIP3-J_0	1	$3 \times K$	α, θ
SIP3-J	K	$3 \times K$	α, θ
SIP4-J_0	3	$3 \times K$	α, θ
SIP4-J	$3 \times K$	$3 \times K$	α, θ

$K \rightarrow$ Total number of iterations.

It is also worth commenting that the s-parameters were only defined in the range [0,1] in [32,33]. However, we have not limited their analysis to this range. Therefore, we have avoided this definition in order to avoid misleading.

3. Convergence Analysis of Studied Newton-SIP Methods

In this section, the convergence rate of the studied Newton-SIP methods is derived. In this case, we consider that the Jacobian matrix is updated each iteration, since the convergence features for iterative procedures (4)–(7) can be derived from this analysis. For this study, the Taylor Expansion technique has been used (see [35] for details).

Theorem 1. Let $\mathbf{g}$ be sufficiently differentiable at each point of an open neighborhood D of $r \in \mathbb{R}^n$, this is a solution of the system $\mathbf{g}(x) = \mathbf{0}$. Let us suppose that $\mathbf{g}(x)$ is continuous and nonsingular in x. Then, the SIP1-J converges to r with the following error function.

$$e^{(k+1)} = \left(1-\alpha^2\right)C_2 e^{(k)^2} + \left(\left(\alpha^3 - 3\alpha^2 + 2\right)C_3 + \left(4\alpha^2 - 2\right)C_2^2\right)e^{(k)^3} + O\left(e^{(k)^4}\right) \quad (8)$$

where $C_j = (1/j!)\left[g'(r)\right]^{-1} g^{(j)}(r)$, $j = 2, 3, \ldots$, and $e^{(k)} = x^{(k)} - r$.

Proof. Taylor expansion of $\mathbf{g}(x)$ and $J(x)$ about r yields:

$$g\left(x^{(k)}\right) = J(r)\left[e^{(k)} + C_2 e^{(k)^2} + C_3 e^{(k)^3}\right] + O\left(e^{(k)^4}\right) \tag{9}$$

$$J\left(x^{(k)}\right) = J(r)\left[I + 2C_2 e^{(k)} + 3C_3 e^{(k)^2}\right] + O\left(e^{(k)^3}\right) \tag{10}$$

where $I \in \mathbb{R}^{n\times n}$ is the identity matrix. Now, let us assume that:

$$\left[J\left(x^{(k)}\right)\right]^{-1} = \left[c_1 I + c_2 e^{(k)} + c_3 e^{(k)^2}\right][J(r)]^{-1} + O\left(e^{(k)^3}\right) \tag{11}$$

where c's$\in \mathbb{R}$. Considering the following inverse definition:

$$\left[J\left(x^{(k)}\right)\right]^{-1} g'\left(x^{(k)}\right) = J\left(x^{(k)}\right)\left[g'\left(x^{(k)}\right)\right]^{-1} = I. \tag{12}$$

By solving the resulting linear system, one can obtain:

$$\left[J\left(x^{(k)}\right)\right]^{-1} = \left[I - 2C_2 e^{(k)} + \left(4C_2^2 - 3C_3\right)e^{(k)^2}\right][J(r)]^{-1} + O\left(e^{(k)^4}\right). \tag{13}$$

Now, let us define $\widetilde{e}^{(k)} = \left(x^{(k)} - \left[J\left(x^{(k)}\right)\right]^{-1} g\left(x^{(k)}\right)\right) - r$ and $e_y^{(k)} = y^{(k)} - r$; thus, one can obtain:

$$e_y^{(k)} = (1-\alpha)e^{(k)} + \alpha\widetilde{e}^{(k)} = (1-\alpha)e^{(k)} + \alpha C_2 e^{(k)^2} + O\left(e^{(k)^3}\right). \tag{14}$$

The Taylor expansion of $\mathbf{g}(y)$ about r yields:

$$g\left(y^{(k)}\right) = J(r)\left[e_y^{(k)} + C_2 e_y^{(k)^2} + C_3 e_y^{(k)^3}\right] + O\left(e_y^{(k)^4}\right). \tag{15}$$

Now, we can calculate the error vector at $k+1$ as follows:

$$e^{(k+1)} = \left(y^{(k)} - \left[J\left(x^{(k)}\right)\right]^{-1} g\left(y^{(k)}\right)\right) - r. \tag{16}$$

After some manipulations, one obtains:

$$e^{(k+1)} = \left(1-\alpha^2\right)C_2 e^{(k)^2} + \left(\left(\alpha^3 - 3\alpha^2 + 2\right)C_3 + \left(4\alpha^2 - 2\right)C_2^2\right)e^{(k)^3} + O\left(e^{(k)^4}\right). \tag{17}$$

The proof is complete. □

Theorem 2. *Let* $\mathbf{g}$ *be sufficiently differentiable at each point of an open neighborhood D of* $r \in \mathbb{R}^n$*; this is a solution of the system* $\mathbf{g}(x) = \mathbf{0}$*. Let us suppose that* $\mathbf{g}(x)$ *is continuous and nonsingular in* x*. Then, the SIP2-J converges to* r *with the following error function.*

$$\begin{aligned} e^{(k+1)} = \; & \left(\alpha^2 - 2\alpha + 1\right)C_2 e^{(k)^2} + \left(\left(6\alpha^2 - 2\alpha^3 - 6\alpha + 2\right)C_3 + \left(2\alpha^3 - 8\alpha^2 + 8\alpha - 2\right)C_2^2\right)e^{(k)^3} \\ & + \left(\left(34\alpha^3 - 7\alpha^4 - 54\alpha^2 + 34\alpha - 7\right)C_2 C_3 + \left(4\alpha^4 - 22\alpha^3 + 37\alpha^2 - 22\alpha + 4\right)C_2^3\right)e^{(k)^4} \\ & + O\left(e^{(k)^5}\right) \end{aligned} \tag{18}$$

Proof. Taking $e_y^{(k)}$ and the Taylor expansion of $\mathbf{g}(y)$ about r from (14) and (15), respectively, let us calculate the Taylor expansion of $[J(y)]^{-1}$ as (12). Thus, one can obtain:

$$\left[J\left(y^{(k)}\right)\right]^{-1} = \left[I - 2C_2 e_y^{(k)} + \left(4C_2^2 - 3C_3\right)e_y^{(k)^2}\right][J(r)]^{-1} + O\left(e_y^{(k)^3}\right). \tag{19}$$

Now, we can calculate the error vector at $k+1$ as follows:

$$e^{(k+1)} = \left(y^{(k)} - \left[J\left(y^{(k)}\right)\right]^{-1} g\left(y^{(k)}\right)\right) - r. \tag{20}$$

After some manipulations, one obtains:

$$e^{(k+1)} = \left(\alpha^2 - 2\alpha + 1\right) C_2 e^{(k)^2} + \left(\left(6\alpha^2 - 2\alpha^3 - 6\alpha + 2\right) C_3 + \left(2\alpha^3 - 8\alpha^2 + 8\alpha - 2\right) C_2^2\right) e^{(k)^3} + \left(\left(34\alpha^3 - 7\alpha^4 - 54\alpha^2 + 34\alpha - 7\right) C_2 C_3 + \left(4\alpha^4 - 22\alpha^3 + 37\alpha^2 - 22\alpha + 4\right) C_2^3\right) e^{(k)^4} + O\left(e^{(k)^5}\right). \tag{21}$$

The proof is complete. □

Theorem 3. *Let* $\mathbf{g}$ *be sufficiently differentiable at each point of an open neighborhood D of* $\boldsymbol{r} \in \mathbb{R}^n$*; this is a solution of the system* $\mathbf{g}(\boldsymbol{x}) = \mathbf{0}$*. Let us suppose that* $\mathbf{g}(\boldsymbol{x})$ *is continuous and nonsingular in* $\boldsymbol{x}$*. Then, the SIP3-J converges to* $\boldsymbol{r}$ *with the following error function.*

$$e^{(k+1)} = \left(2 - 2\alpha\theta^2\right) C_2^2 e^{(k)^3} + \left(\left(3 + 2\alpha\theta^3 - 9\alpha\theta^2 + 4\alpha\right) C_2 C_3 + \left(-\alpha^2\theta^4 + 14\alpha\theta^2 - 4\alpha - 5\right) C_2^3\right) e^{(k)^4} + O\left(e^{(k)^5}\right) \tag{22}$$

Proof. Let us define $\tilde{e}^{(k)} = \left(x^{(k)} - \left[J\left(x^{(k)}\right)\right]^{-1} \mathbf{g}\left(x^{(k)}\right)\right) - r$ and $e_z^{(k)} = z^{(k)} - r$; then, one can obtain:

$$e_z^{(k)} = (1 - \theta) e^{(k)} + \theta \tilde{e}^{(k)} = (1 - \theta) e^{(k)} + \theta C_2 e^{(k)^2} + O\left(e^{(k)^3}\right). \tag{23}$$

The Taylor expansion of $\mathbf{g}(z)$ about r yields:

$$\mathbf{g}\left(z^{(k)}\right) = J(r)\left[e_z^{(k)} + C_2 e_z^{(k)^2} + C_3 e_z^{(k)^3}\right] + O\left(e_z^{(k)^4}\right). \tag{24}$$

Now, let us define $\hat{e}^{(k)} = \left(z^{(k)} - \left[J\left(x^{(k)}\right)\right]^{-1} \mathbf{g}\left(z^{(k)}\right)\right) - r$ and $e_y^{(k)} = y^{(k)} - r$; then, we can calculate:

$$e_y^{(k)} = (1 - \alpha)\tilde{e}^{(k)} + \alpha \hat{e}^{(k)} = \left(1 - \alpha\theta^2\right) C_2 e^{(k)^2} + \left(\left(4\alpha\theta^2 - 2\alpha\right) C_2^2 + \left(\alpha\theta^3 - 3\alpha\theta^2 + 2\alpha\right) C_3\right) e^{(k)^3} + \left(\left(4\alpha - 9\alpha\theta^2\right) C_2^3 + \left(15\alpha\theta^2 - 5\alpha\theta^3 - 7\alpha\right) C_2 C_3\right) e^{(k)^4} + O\left(e^{(k)^5}\right). \tag{25}$$

The Taylor expansion of $\mathbf{g}(y)$ about r yields:

$$\mathbf{g}\left(y^{(k)}\right) = J(r)\left[e_y^{(k)} + C_2 e_y^{(k)^2} + C_3 e_y^{(k)^3}\right] + O\left(e_y^{(k)^4}\right). \tag{26}$$

Now, we can calculate the error vector at $k + 1$ as follows:

$$e^{(k+1)} = \left(y^{(k)} - \left[J\left(x^{(k)}\right)\right]^{-1} \mathbf{g}\left(y^{(k)}\right)\right) - r. \tag{27}$$

After some manipulations, one can obtain:

$$e^{(k+1)} = \left(2 - 2\alpha\theta^2\right) C_2^2 e^{(k)^3} + \left(\left(3 + 2\alpha\theta^3 - 9\alpha\theta^2 + 4\alpha\right) C_2 C_3 + \left(-\alpha^2\theta^4 + 14\alpha\theta^2 - 4\alpha - 5\right) C_2^3\right) e^{(k)^4} + O\left(e^{(k)^5}\right). \tag{28}$$

The proof is complete. □

Theorem 4. *Let* $\mathbf{g}$ *be sufficiently differentiable at each point of an open neighborhood D of* $\boldsymbol{r} \in \mathbb{R}^n$*; this is a solution of the system* $\mathbf{g}(\boldsymbol{x}) = 0$*. Let us suppose that* $\mathbf{g}(\boldsymbol{x})$ *is continuous and nonsingular in* $\boldsymbol{x}$*. Then, the SIP4-J converges to* $\boldsymbol{r}$ *with the following error function.*

$$e^{(k+1)} = X_4 e^{(k)^4} + X_5 e^{(k)^5} + X_6 e^{(k)^6} + X_7 e^{(k)^7} + X_8 e^{(k)^8} + O\left(e^{(k)^9}\right) \tag{29}$$

where:

$$X_4 = \left(\alpha^2\theta^4 - 4\alpha^2\theta^3 + 4\alpha^2\theta^2 + 2\alpha\theta^2 - 4\alpha\theta\right) C_2^3 \tag{30}$$

$$X_5 = \left(20\alpha^2\theta^4 - 4\alpha^2\theta^5 - 36\alpha^2\theta^3 + 28\alpha^2\theta^2 - 4\alpha\theta^3 - 8\alpha^2\theta + 12\alpha\theta^2 - 12\alpha\theta + 4\alpha\right) C_2^2 C_3 + \left(4\alpha^2\theta^5 - 24\alpha^2\theta^4 + 48\alpha^2\theta^3 - 36\alpha^2\theta^2 + 4\alpha\theta^3 + 8\alpha^2\theta - 16\alpha\theta^2 + 16\alpha\theta - 4\alpha\right) C_2^4 \tag{31}$$

$$
\begin{aligned}
\mathbf{X_6} = \big(4\alpha^2\theta^6 \; & -24\alpha^2\theta^5 + 60\alpha^2\theta^4 - 80\alpha^2\theta^3 + 60\alpha^2\theta^2 - 24\alpha^2\theta + 4\alpha^2\big)C_2C_3^2 \\
& + \big(+2\alpha^3\theta^6 - 12\alpha^3\theta^5 - 8\alpha^2\theta^6 + 24\alpha^3\theta^4 + 56\alpha^2\theta^5 - 16\alpha^3\theta^3 - 146\alpha^2\theta^4 + 184\alpha^2\theta^3 \\
& -128\alpha^2\theta^2 + 56\alpha^2\theta + 6\alpha\theta^2 - 8\alpha^2 - 12\alpha\theta + 2\big)C_2^3C_3 \\
& + \big(12\alpha^3\theta^5 - 2\alpha^3\theta^6 + 4\alpha^2\theta^6 - 24\alpha^3\theta^4 - 32\alpha^2\theta^5 + 16\alpha^3\theta^3 + 90\alpha^2\theta^4 - 112\alpha^2\theta^3 \\
& +72\alpha^2\theta^2 - 32\alpha^2\theta - 6\alpha\theta^2 + 4\alpha^2 + 12\alpha\theta - 2\big)C_2^5
\end{aligned} \tag{32}
$$

$$
\begin{aligned}
\mathbf{X_7} = \big(96\alpha^3\theta^6 \; & -12\alpha^3\theta^7 - 288\alpha^3\theta^5 + 396\alpha^3\theta^4 - 24\alpha^2\theta^5 - 240\alpha^3\theta^3 + 144\alpha^2\theta^4 + 48\alpha^3\theta^2 - 288\alpha^2\theta^3 \\
& +216\alpha^2\theta^2 - 12\alpha\theta^3 - 48\alpha^2\theta + 48\alpha\theta^2 - 48\alpha\theta + 12\alpha\big)C_2^6 \\
& + \big(24\alpha^3\theta^7 - 180\alpha^3\theta^6 + 516\alpha^3\theta^5 - 696\alpha^3\theta^4 + 48\alpha^2\theta^5 + 432\alpha^3\theta^3 - 264\alpha^2\theta^4 - 96\alpha^3\theta^2 \\
& +504\alpha^2\theta^3 - 384\alpha^2\theta^2 + 24\alpha\theta^3 + 96\alpha^2\theta - 84\alpha\theta^2 + 84\alpha\theta - 24\alpha\big)C_2^4C_3 \\
& + \big(-12\alpha^3\theta^7 + 84\alpha^3\theta^6 - 228\alpha^3\theta^5 + 300\alpha^3\theta^4 - 24\alpha^2\theta^5 - 192\alpha^3\theta^3 + 120\alpha^2\theta^4 + 48\alpha^3\theta^2 \\
& -216\alpha^2\theta^3 + 168\alpha^2\theta^2 - 12\alpha\theta^3 - 48\alpha^2\theta + 36\alpha\theta^2 - 36\alpha\theta + 12\alpha\big)C_2^2C_3^2
\end{aligned} \tag{33}
$$

$$
\begin{gathered}
\mathbf{X_8} = \big(4\alpha^4\theta^8 - 32\alpha^4\theta^7 - 24\alpha^3\theta^8 + 96\alpha^4\theta^6 + 240\alpha^3\theta^7 - 16\alpha^2\theta^8 - 128\alpha^4\theta^5 - 944\alpha^3\theta^6 + 176\alpha^2\theta^7 + \\
64\alpha^4\theta^4 + 1872\alpha^3\theta^5 - 804\alpha^2\theta^6 - 2016\alpha^3\theta^4 + 1996\alpha^2\theta^5 + 1216\alpha^3\theta^3 - 2921\alpha^2\theta^4 - 408\alpha^3\theta^2 + \\
2524\alpha^2\theta^3 + 48\alpha^3\theta - 1260\alpha^2\theta^2 + 368\alpha^2\theta + 16\alpha\theta^2 - 40\alpha^2 - 32\theta\alpha + 4\big)C_2^7 + \big(-7\alpha^4\theta^8 + 56\alpha^4\theta^6 + \\
72\alpha^3\theta^8 - 168\alpha^4\theta^6 - 672\alpha^3\theta^7 + 56\alpha^2\theta^8 + 224\alpha^4\theta^5 + 2516\alpha^3\theta^6 - 580\alpha^2\theta^7 - 112\alpha^4\theta^4 - 4872\alpha^3\theta^5 + \\
2518\alpha^2\theta^6 + 5280\alpha^3\theta^4 - 6000\alpha^2\theta^5 - 3280\alpha^3\theta^3 + 8546\alpha^2\theta^4 + 1128\alpha^3\theta^2 - 7368\alpha^2\theta^3 - 144\alpha^3\theta + \\
3766\alpha^2\theta^2 - 1108\alpha^2\theta - 28\alpha\theta^2 + 128\alpha^2 + 56\alpha\theta - 7\big)C_2^5C_3 + \big(624\alpha^3\theta^7 - 49\alpha^2\theta^8 - 2232\alpha^3\theta^6 + 476\alpha^2\theta^7 + \\
4272\alpha^3\theta^5 - 1984\alpha^2\theta^6 - 4728\alpha^3\theta^4 + 4628\alpha^2\theta^5 - 72\alpha^3\theta^8 + 3024\alpha^3\theta^3 - 6598\alpha^2\theta^4 - 1032\alpha^3\theta^2 + \\
5876\alpha^2\theta^3 + 144\alpha^3\theta - 3184\alpha^2\theta^2 + 956\alpha^2\theta - 121\alpha^2\big)C_2^3C_3^2 + \big(24\alpha^3\theta^8 - 192\alpha^3\theta^7 + 648\alpha^3\theta^6 - \\
1200\alpha^3\theta^5 + 24\alpha^2\theta^6 + 1320\alpha^3\theta^4 - 144\alpha^2\theta^5 - 864\alpha^3\theta^3 + 360\alpha^2\theta^4 + 312\alpha^3\theta - 480\alpha^2\theta^3 - 48\alpha^3\theta + \\
360\alpha^2\theta^2 - 144\alpha^2\theta + 24\alpha^2\big)C_2C_3^3
\end{gathered} \tag{34}
$$

Proof. In this case, one can calculate the Taylor expansion of $[J(\boldsymbol{y})]^{-1}$ about $\boldsymbol{r}$ as (12). Thus, one can obtain:

$$\left[J\left(z^{(k)}\right)\right]^{-1} = \left[I - 2C_2 e_z^{(k)} + \left(4C_2^2 - 3C_3\right) e_z^{(k)^2}\right] [J(\boldsymbol{r})]^{-1} + O\left(e_z^{(k)^3}\right) \tag{35}$$

where $e_z^{(k)}$ is already calculated in (23). Now, taking the Taylor expansion of $\mathbf{g}(z)$ about $\boldsymbol{r}$ from (24), let us define $\hat{e}^{(k)} = \left(z^{(k)} - \left[J\left(z^{(k)}\right)\right]^{-1} \mathbf{g}\left(z^{(k)}\right)\right) - \boldsymbol{r}$ and $e_y^{(k)} = \boldsymbol{y}^{(k)} - \boldsymbol{r}$; then, we can calculate:

$$e_y^{(k)} = (1-\alpha)\widetilde{e}^{(k)} + \alpha \hat{e}^{(k)} \tag{36}$$

where $\widetilde{e}^{(k)} = \left(\boldsymbol{x}^{(k)} - \left[J\left(\boldsymbol{x}^{(k)}\right)\right]^{-1} \mathbf{g}\left(\boldsymbol{x}^{(k)}\right)\right) - \boldsymbol{r}$. Now, manipulating (31), one obtains:

$$e_y^{(k)} = \left(\alpha\theta^2 - 2\alpha\theta + 1\right) C_2 e^{(k)^2} + \left(\left(2\alpha\theta^3 - 8\alpha\theta^2 + 8\alpha\theta - 2\alpha\right) C_2^2 + \left(6\alpha\theta^2 - 2\alpha\theta^3 - 6\alpha\theta + 2\alpha\right) C_3\right) e^{(k)^3} + O\left(e^{(k)^4}\right). \tag{37}$$

The Taylor expansion of $\mathbf{g}(\boldsymbol{y})$ and of $[J(\boldsymbol{y})]^{-1}$ about $\boldsymbol{r}$ yields:

$$\mathbf{g}\left(\boldsymbol{y}^{(k)}\right) = J(\boldsymbol{r}) \left[e_y^{(k)} + C_2 e_y^{(k)^2} + C_3 e_y^{(k)^3}\right] + O\left(e_y^{(k)^4}\right) \tag{38}$$

$$\left[J\left(\boldsymbol{y}^{(k)}\right)\right]^{-1} = \left[I - 2C_2 e_y^{(k)} + \left(4C_2^2 - 3C_3\right) e_y^{(k)^2}\right] [J(\boldsymbol{r})]^{-1} + O\left(e_y^{(k)^3}\right). \tag{39}$$

At this point, we can calculate the error vector at $k+1$ as follows:

$$e^{(k+1)} = \left(\boldsymbol{y}^{(k)} - \left[J\left(\boldsymbol{x}^{(k)}\right)\right]^{-1} \mathbf{g}\left(\boldsymbol{y}^{(k)}\right)\right) - \boldsymbol{r}. \tag{40}$$

After some manipulations, one can obtain:

$$e^{(k+1)} = \mathbf{X_4} e^{(k)^4} + \mathbf{X_5} e^{(k)^5} + \mathbf{X_6} e^{(k)^6} + \mathbf{X_7} e^{(k)^7} + \mathbf{X_8} e^{(k)^8} + O\left(e^{(k)^9}\right) \tag{41}$$

where the **X**s are defined in (30)–(34); hence, the proof is complete. □

At this point, one can easily check that the studied techniques achieve their highest convergence rates when the *s*-parameters are equal to 1. Thus, SIP1-J, SIP2-J, SIP3-J, and SIP4-J can show third, fourth, fourth, and eighth convergence rates, respectively. In this regard, it is important to note that the mentioned techniques present a higher convergence rate compared with NR, which presents quadratic convergence. On the other hand, it is worth mentioning that the studied Newton-SIP methods have the characteristic lowest convergence rate even when α, $\theta \neq 1$. By observing the error functions of these techniques, one can deduce that the order of convergence of SIP1-J and SIP2-J is at least two, while it is three in the case of SIP3-J and four for SIP4-J.

In the case of the standard forms of studied Newton-SIP methods (Equations (4)–(7)), since their convergence rates do not remain constant during the iterative procedure, it is more suitable to study them by successive substitution and Taylor expansion in their respective algorithms (details of this technique can be found in [36]). Figure 1 shows the convergence rates of the algorithms SIP1-J_0, SIP2-J_0, SIP3-J_0, and SIP4-J_0 as a function of the iteration number. In this figure, the convergence rate of the conventional NR has been also included for comparison. From this figure, it can be deduced that SIP4-J_0 and SIP1-J_0 show the highest and the lowest convergence rate, respectively, while both SIP2-J_0 and SIP3-J_0 have the same convergence order. Anyway, for these techniques, the convergence rate is linear after the first iteration. Therefore, their convergence orders are always less than two, being so overcome by NR.

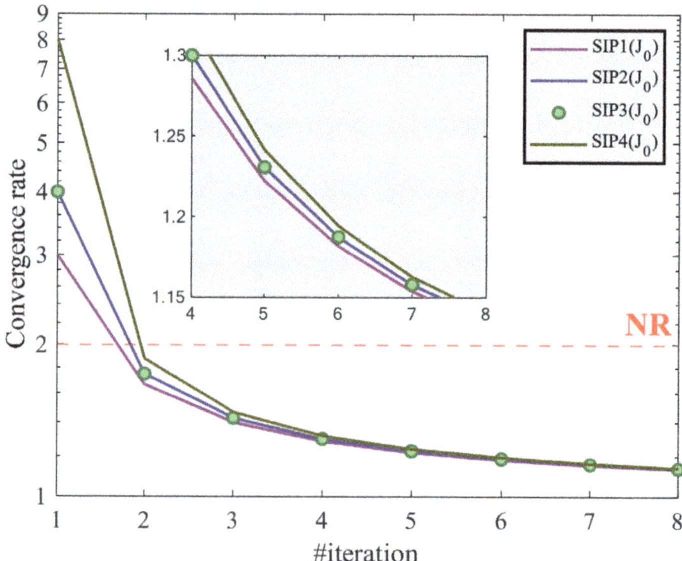

Figure 1. Convergence rates of the algorithms (4)–(7).

To compare the algorithms (4)–(7) and their respective counterparts (in which the Jacobian is updated each iteration), let us refer to the convergence rate of the initial error vector. It means, let us suppose that the error vector evolves as $e^{(0)^j}$. Thus, as j grows, the solution is assumed to be more closely approached. Therefore, it is assumed that an algorithm will converge faster as j grows rapidly. Figure 2 plots the value of j of the studied Newton-SIP methods at different iterations. From this figure, it can be seen that j exponentially grows in the case of SIP1-J, SIP2-J, SIP3-J, and SIP4-J, while it grows linearly with SIP1-J_0 SIP2-J_0, SIP3-J_0, and SIP4-J_0. To complete the section, Table 2 summarizes the convergence analysis of the considered techniques and the NR. As commented, the studied techniques achieve their maximum convergence rate when the *s*-parameters are equal to 1.

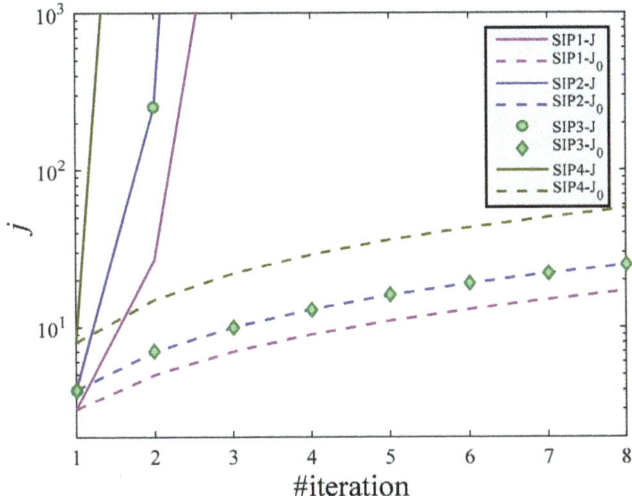

Figure 2. Convergence degree of the error vector $e^{(0)}$ at different iteration counters.

Table 2. Comparison of the convergence rates of studied PF solution techniques.

Method	Convergence Rate	
	Minimum	Maximum
NR	2	2
SIP1-J0	Linear	Linear
SIP1-J	2	3
SIP2-J0	Linear	Linear
SIP2-J	3	4
SIP3-J0	Linear	Linear
SIP3-J	3	4
SIP4-J0	Linear	Linear
SIP4-J	4	8

4. Comparison of the Efficiency of Different Iterative Algorithms

In this section, we compare the efficiency of the studied PF solution methods. To do that, let us consider the following efficiency index [37].

$$FEI = p^{\frac{1}{CO}} \tag{42}$$

where $p \in \mathbb{R}^+$ is the order of convergence, and CO stands for the total computational cost of an iteration. In this sense, the following theorem is generally used to estimate the cost of a LU decomposition [35].

Theorem 5. The number of products and quotients required for solving q linear systems of equations with the same matrix of coefficients, using LU factorization, is:

$$o(n,q) = \frac{1}{3}n^3 + qn^2 - \frac{1}{3}n. \tag{43}$$

It is also suitable to consider the computational cost of each function evaluation $o(\mathbf{g})$ and Jacobian evaluation $o\left(\mathbf{g}'\right)$; hence, the total computational cost of an iteration can be estimated as follows:

$$CO = o(\mathbf{g}) + o\left(\mathbf{g}'\right) + o(n,q). \tag{44}$$

The value of the index (42) for studied Newton-SIP methods and NR is depicted in Figure 3 for different sizes of the PF state vector (n). From this figure, it can be appreciated that Newton-SIP techniques are more efficient than NR. In addition, the following relations hold:

$$FEI_{\text{NR}} < FEI_{\text{SIP2-J}} = FEI_{\text{SIP4-J}} < FEI_{\text{SIP1-J}} < FEI_{\text{SIP3-J}}. \quad (45)$$

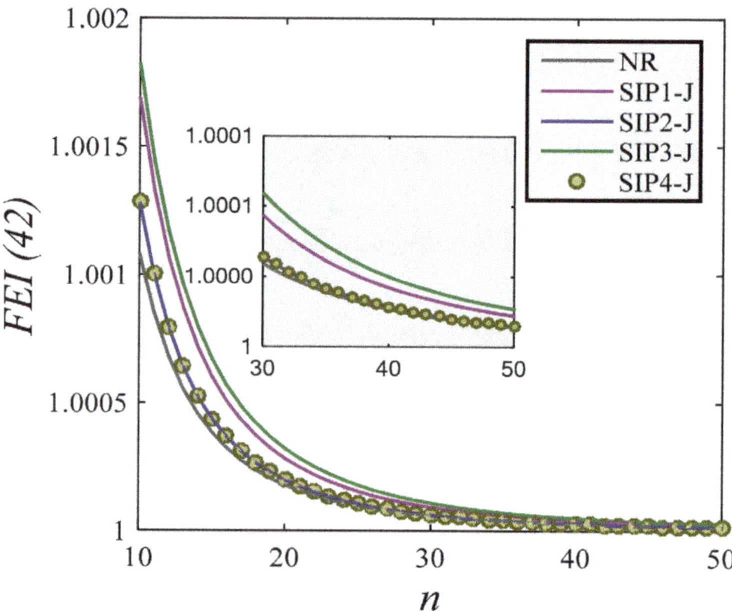

Figure 3. Efficiency index (42) for different PF solvers.

5. Numerical Experiments

In this section, several numerical experiments are carried out, and their results are analyzed. The studied Newton-SIP PF techniques are compared with NR in several test systems. All simulations have been done using Matpower v7.0 [38]. The studied systems have been taken from MATPOWER's database [39–41].

In all simulations, $\mathbf{g}(\boldsymbol{x})_\infty \leq 10^{-6}$ has been taken as a convergence criterion, and a flat start has been considered for initializing the PF analysis. The reported execution times have been obtained under Windows 10 on a 3.4 GHz Intel Core i7-8750H CPU 2.2 GHz personal laptop (16.00 GB RAM) and calculated as the average value of 1000 simulations.

5.1. Well-Conditioned Cases

Firstly, we have analyzed the performance of studied PF techniques in several well-conditioned systems, which range from 30 to 3120 buses. Figure 4 shows the obtained results of these systems. These results have been obtained for the maximum convergence rate of the studied Newton-SIP PF techniques, i.e., when the *s*-parameters are 1.

As expected, SIP1-J_0 and SIP3-J_0 are the fastest methods, which is strongly linked with the number of factorizations required (one should note that the LU decomposition is the heaviest part of any PF calculation [13]). Among all the studied Newton-SIP techniques, only SIP2-J and SIP4-J are occasionally slower than NR. These results may look not coherent with the analysis performed in Section 4; however, Figure 4c provides a clear explanation about this issue. In this figure, it can be appreciated that these two techniques frequently required more factorizations than NR, which is reflected in a higher computational burden and therefore less competitive execution times. Regarding the total iterations required to attain the solution, the results are expected since the highest convergence rate has the lowest number of total iterations required for achieving the solution. Regarding those algorithms with linear convergence (Equations (4)–(7)), the following relations normally hold:

$$it_{\text{SIP4}-\text{J}_0} < it_{\text{SIP2}-\text{J}_0} < it_{\text{SIP3}-\text{J}_0} < it_{\text{SIP1}-\text{J}_0} \quad (46)$$

where it_a indicates the total number of iterations of the method "a". One should note that relations (46) are coherent with the theoretical analysis performed in Section 3 (see Figure 2); nevertheless, there are exceptions such as case2746wop.

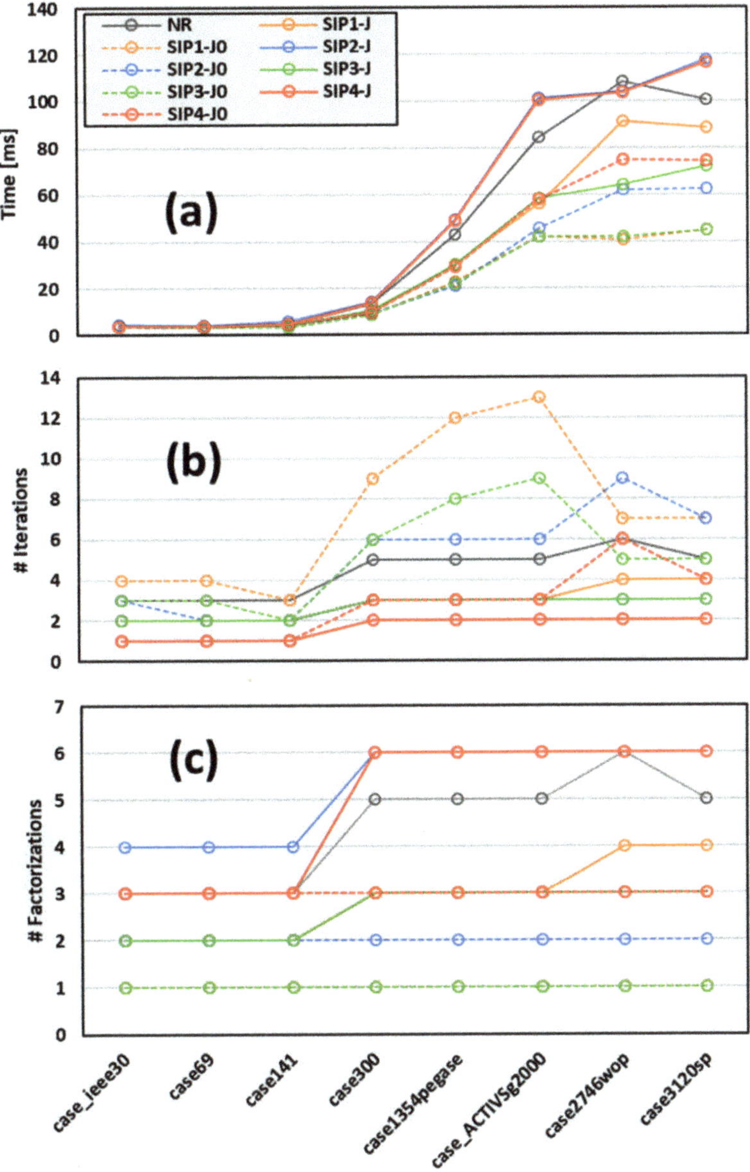

Figure 4. Comparison of the results obtained in well-conditioned systems. (**a**) Execution time (ms), (**b**) total iterations, and (**c**) total factorizations.

Now, let us consider the influence of the loading level. To do that, the active and reactive powers injected at PQ buses along the active powers injected at PV buses have been progressively increased in steps of $\lambda = 0.0001$ pu until all studied techniques have diverged. Figure 5 is analogous to Figure 4 for limit-loading cases. In this case, all techniques diverged for the same loading level; hence, all of them have been tested for the same conditions.

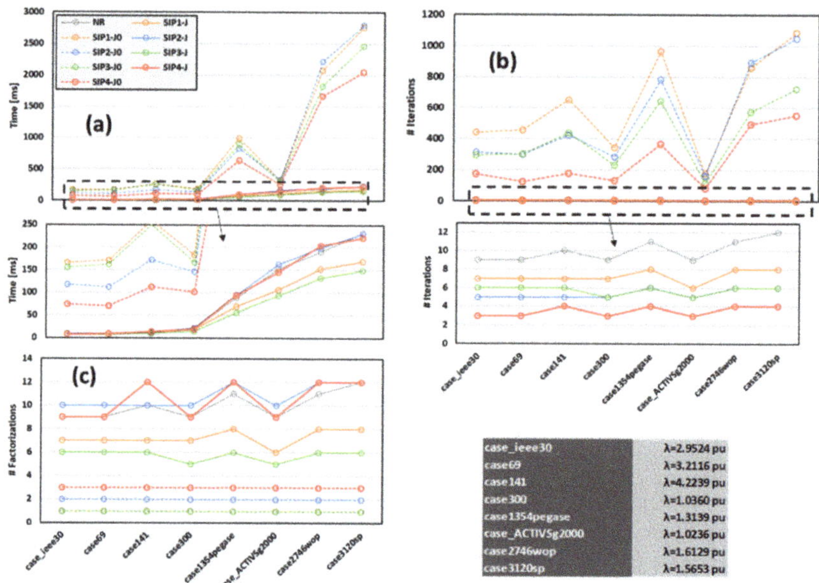

Figure 5. Comparison of the results obtained in well-conditioned systems in a limit load scenario. (**a**) Execution time (ms), (**b**) total iterations number and (**c**) total factorizations.

While similar conclusions can be extracted for SIP1-J, SIP2-J, SIP3-J, and SIP4-J, linear algorithms are not competitive in this scenario due to the huge amount of iterations required to reach the solution. Hence, although they frequently employed very few factorizations, their execution times are not competitive at all. In addition, relations (46) are not held in this situation for SIP2-J_0 and SIP3-J_0, since it can be observed in Figure 5 that clearly $it_{\mathrm{SIP3-J_0}} < it_{\mathrm{SIP2-J_0}}$.

In order to overcome the important drawbacks shown by linear techniques when the loading level is high, s-parameters can be occasionally set greater than one. In this case, these parameters have an accelerating effect on the convergence performance of the studied Newton-SIP techniques. Figure 6 shows the total number of iterations of different linear Newton-SIP techniques when their s-parameters are fixed greater than one and therefore achieve the fastest convergence. Under these settings, the results are compared with those results depicted in Figure 5. In this case, s-parameters have been obtained empirically. As can be seen, the total number of iterations can be drastically reduced; however, the criteria for determining the best values of s-parameters do not follow any specific pattern, as shown in Figure 6b. In light of the results obtained, this topic looks strongly case-dependent.

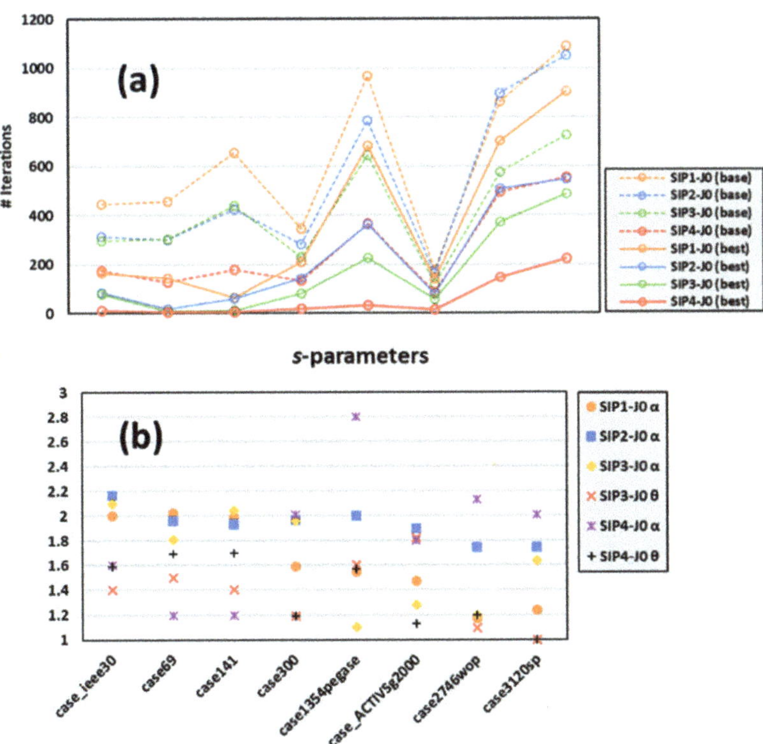

Figure 6. (**a**) Comparison of the total number of iterations of different linear Newton-SIP solvers, taking the *s*-parameters equal to 1. (**b**) Considered *s*-parameters.

5.2. Ill-Conditioned Cases

Now, we have considered case3012wp, case3375wp, and case13659pegase, which are available in MATPOWER's database. These cases correspond with snapshots of real cases, which demonstrates that ill-conditioned solvers may appear in real applications and they are even more frequent nowadays [42]. NR fails to solve these systems when a flat start is used, so that they can be categorized as ill-conditioned [13]. In previous simulations, we have taken the *s*-parameters to be equal to one, since the studied methods achieve their maximum convergence rate for these values. However, in ill-conditioned systems, this strategy may lead to divergence. Therefore, it is more suitable to study which values of the *s*-parameters in the considered Newton-SIP methods are reliable. Figures 7–9 show the areas of successful convergence for the studied ill-conditioned cases. Hence, it is considered to have failed since both divergence and convergence lead to inaccurate solutions.

Firstly, it can be easily appreciated that SIP1-J, SIP2-J, SIP3-J, and SIP4-J are typically less reliable than SIP1-J_0, SIP2-J_0, SIP3-J_0, and SIP4-J_0, since the latter normally showed wider convergence areas. There are some remarkable cases; for example, SIP1-J did not converge in the case3012wp and case3375wp; on the other hand, SIP1-J_0 and SIP3-J_0 frequently converged, regardless of the value of parameters. Finally, SIP4-J and SIP4-J_0 look very sensitive to the values of *s*-parameters. Their convergence rate precisely explains the superior robustness features of linear methods. In [43], it is said that the methods with high convergence rates normally show narrow Regions of Attraction; in other words, the highest convergence rate has the most sensitivity with respect to the initial guess. This fact can also be appreciated for other PF techniques such as [10,14], which introduce a discrete step size to reduce the convergence rate and obtain robust techniques properly.

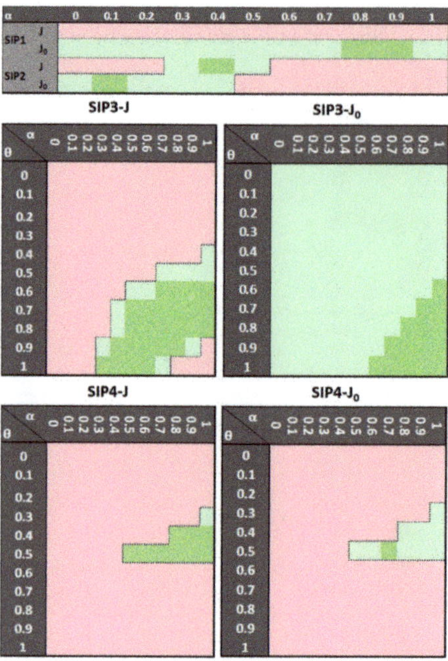

Figure 7. Convergence (green) and failure areas (red) in the s-parameter space for solving the case3012wp. Bold green areas indicate where the studied technique successfully converged, employing the least number of iterations.

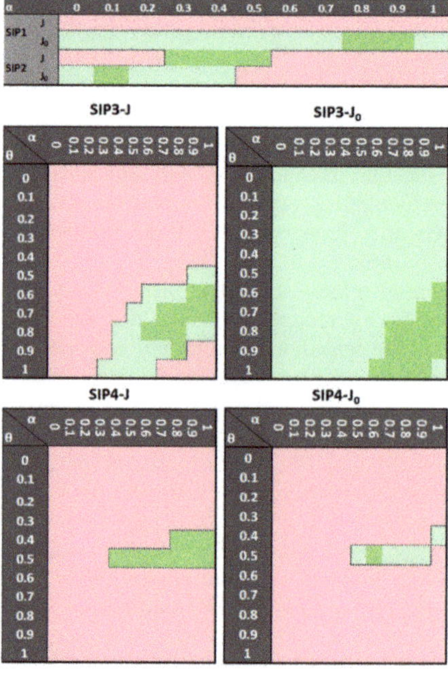

Figure 8. Convergence (green) and failure areas (red) in the s-parameter space for solving the case3375wp. Bold green areas indicate where the studied technique successfully converged, employing the least number of iterations.

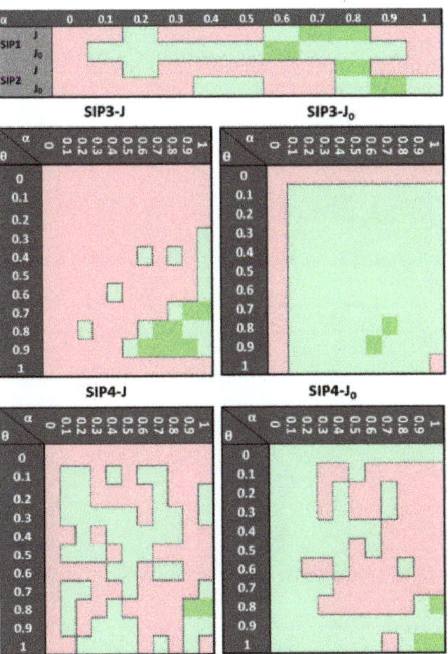

Figure 9. Convergence (green) and failure areas (red) in the s-parameter space for solving the case13659pegase. Bold green areas indicate where the studied technique successfully converged, employing the least number of iterations.

From Figures 7–9, it can also be seen that tuning the s-parameters is a strongly case-dependent topic. For example, while the considered techniques performed very similar in case3012wp and case3375wp, the behavior of the techniques became erratic in the case13659pegase. In addition, the least number of iterations is not always achieved for the same s-parameters.

In order to properly compare the studied Newton-SIP techniques in these ill-conditioned cases, let us consider only those cases in which the considered techniques required the least number of iterations for successfully converging. Figure 10 shows the execution time along with the total number of iterations and factorizations employed by the Newton-SIP PF techniques in the studied ill-conditioned systems. As commented, NR failed in these cases.

As in well-conditioned systems, SIP1-J_0, SIP2-J_0, and SIP3-J_0 are the fastest techniques, while SIP1-J, SIP2-J, SIP3-J, and SIP4-J typically reached the solution employing less iterations. It is worth mentioning that SIP3-J is occasionally faster than SIP4-J_0 in the case13659pegase. This is because these two methods require the same number of factorizations in this system; however, SIP4-J_0 computes more calculations per iteration. SIP1-J only converged in the case13659pegase.

To conclude this analysis, the influence of the initial guess x_0 on the convergence features of each studied technique has been analyzed. To this end, two cases have been compared. On the one hand, we assume a flat start as in previous simulations, taking the best value of the *s*-parameters for each case. On the other hand, we took the default starter provided in Matpower, which is normally closer to the solution than the flat initialization. Figure 11 shows the number of iterations for these two cases in the case3012wp. As observed, the quality of the initialization directly affects the convergence of the studied methods, normally employing more iterations when a flat start is used. These results are coherent, since the closer the solution is to the start point, the faster the convergence. This same conclusion was attained in other recent papers such as [44].

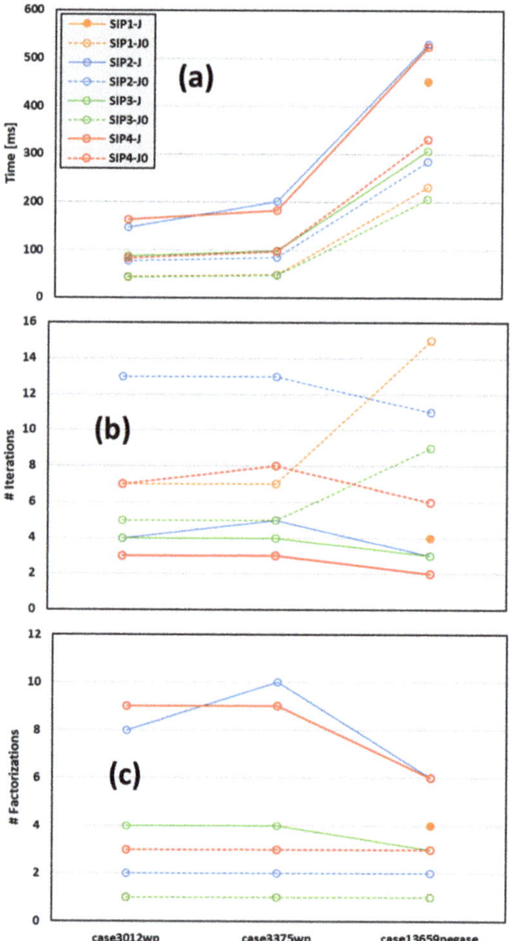

Figure 10. Comparison of the results obtained in ill-conditioned systems. (**a**) Execution time (ms), (**b**) total iterations number, and (**c**) total factorizations. *S*-parameters have been tuned as all studied techniques employed the least number of iterations.

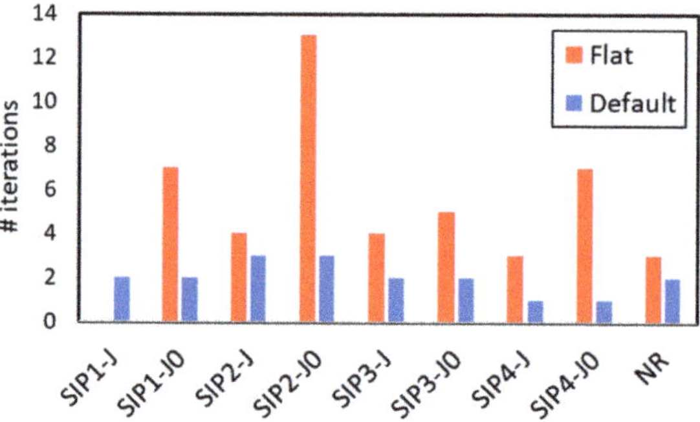

Figure 11. Total iterations employed in case3012wp with different solvers considering a flat start and the default starter provided in Matpower.

It is also worth noting how the starter point affects the robustness of the solvers. Particularly, in the case of SIP1-J, the solution is successfully achieves starting from the default point, while this solver failed from the flat starter. This is coherent with the definition of ill-conditioned systems provided in [13], which is strongly related with the quality of the starting point.

5.3. Influence of the R/X Ratio

It is well-known that the R/X ratio may negatively impact on the convergence features of PF solvers [21]. In this section, we analyze how the studied techniques are affected by this parameter. To this end, we have considered the small-scale case_ieee30, since high R/X ratios are more frequently in this kind of network. To properly analyze this aspect, we have considered that the branch resistances are multiplied by a real factor μ, and the number of iterations for various values of μ is compared in Figure 12. In this case, the s-parameters were fixed equal to 1 in order to attain the highest convergence rate. As expected, the higher the R/X ratio, the higher the number of iterations employed to converge. This fact is especially remarkable in those solvers with fixed Jacobian, provoking divergence in some cases.

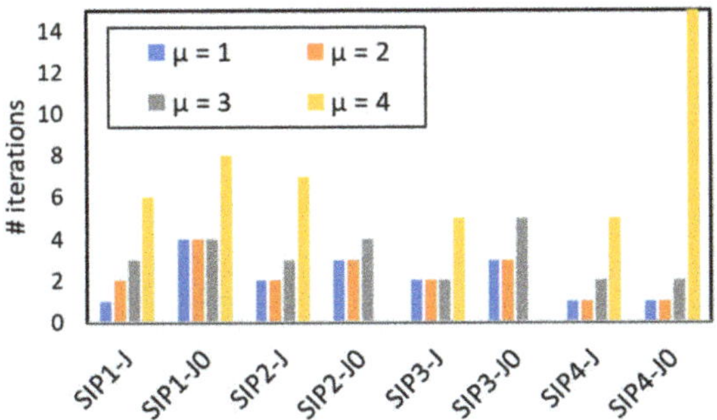

Figure 12. Total iterations employed in case_ieee30 with different solvers and various R/X ratios.

Lastly, we analyze the performance of the developed solvers on a real radial distribution system. To this end, we have considered case18 (18-bus radial distribution system from Grady, Samotyj, and Noyola) and case141 (141-bus radial distribution system from Khodr, Olsina, De Jesus, and Yusta) from Matpower's database. In this case, the results obtained by the studied solvers have been compared with the Forward–Backward sweep algorithm (FBS) [45], which is frequently considered the most conventional solver for radial distribution systems. Figure 13 shows the total number of iterations in the studied radial systems. As observed, the Newton-SIP methods normally outperformed FBS, thus proving their efficacy to handle a wide variety of networks with different features and topologies.

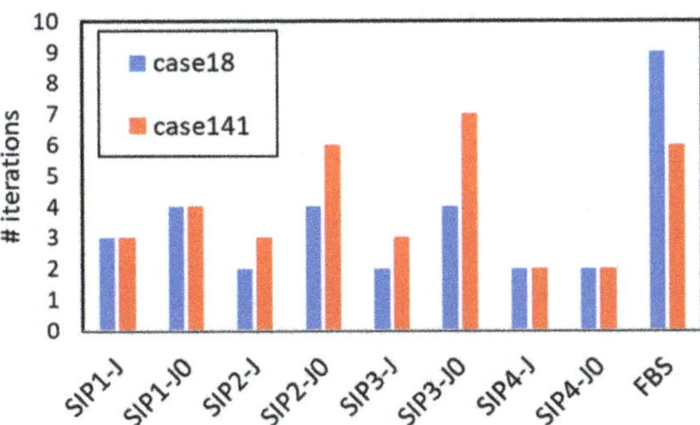

Figure 13. Total iterations employed in the various radial distributions test networks.

6. Conclusions and Future Works

In this paper, the applicability of different Newton-SIP methods for solving the PF problems has been comprehensively studied. Four techniques [32,33] have been considered, and two different iterative schemes have been analyzed. Firstly, we have considered the standard form of Newton-SIP techniques, i.e., those algorithms in which the Jacobian matrix is only updated at the first iteration. Secondly, a fully updated iterative scheme has been considered, in which the Jacobian matrix is updated each iteration as NR.

The convergence characteristics of the considered PF techniques have been studied. In case a fully updated iterative scheme is considered, the techniques can achieve up to the eighth order of convergence. It has been also demonstrated that the highest convergence rate is achieved when the *s*-parameters are set equal to one. On the other hand, the convergence rate of the standard form of Newton-SIP methods is always linear.

The efficiency of the studied techniques has been compared using a well-known efficiency index. The results indicate that the SIP3-J is the most efficient Newton-SIP technique, since it is able to achieve the 4th order of convergence by only factorizing one Jacobian matrix for each iteration. Nevertheless, all the studied techniques showed higher efficiency indices than NR.

Various numerical experiments have been carried out for several well and ill-conditioned systems with different sizes and topologies. The most remarkable conclusion is the Newton-SIP techniques' ability to manage both well and ill-conditioned systems efficiently. They typically outperformed NR in well-conditioned systems. In addition, they are more robust than NR in ill-conditioned cases. These features make Newton-SIP techniques very suitable for widespread industrial applications, as it was pointed out in [31]. Drawbacks showed by the linear Newton-SIP techniques in heavy loading cases can be overcome using their fully updated schemes. Comparing the studied Newton-SIP techniques, the best trade-off between robustness and efficiency is normally obtained with SIP3-J.

However, the performance of Newton-SIP techniques is notably influenced by the values of the involved *s*-parameters. For example, in heavy loading systems, it has been shown that the convergence characteristics can be notably improved by taking advantage of the accelerating effect of *s*-parameters. On the other hand, the robustness properties of the Newton-SIP techniques are strongly affected by the parameters involved. However, the analysis carried out in this work shows that tuning the *s*-parameters is a strong case-dependent topic. It also looks very difficult to be tackled, since any common pattern has been observed.

In radial distribution systems, the studied techniques outperformed FBS, thus demonstrating their ability to handle a wide variety of networks. These results along those obtained in large-scale well and ill-conditioned cases manifest the suitability of the Newton-SIP methods to be applied in real industry tools and even for voltage stability analysis and optimization problems. Further results will be obtained in future works to confirm that point.

Consequently, future works should be focused on further tackling alternative schemes for optimally tuning the *s*-parameters in order to get a good trade-off between efficiency and robustness and avoid the necessity to be initially set by the user. In this sense, optimal conditions should be derived, thus allowing implementing auxiliary routines to properly set those parameters.

Author Contributions: Conceptualization, M.T.-V.; Data curation, S.K. and I.B.M.T.; Formal analysis, M.T.-V., S.K., I.B.M.T. and F.J.; Funding acquisition, S.K. and I.B.M.T.; Investigation, M.T.-V., S.K. and F.J.; Methodology, M.T.-V., S.K. and F.J.; Project administration, S.K., I.B.M.T. and F.J.; Resources, F.J.; Software, F.J.; Supervision, S.K., I.B.M.T. and F.J.; Validation, S.K.; Visualization, S.K., I.B.M.T. and F.J.; Writing—original draft, M.T.-V. and S.K.; Writing—review and editing, I.B.M.T. and F.J. All authors have read and agreed to the published version of the manuscript.

Funding: This research was funded by Taif University Researchers Supporting Project number (TURSP-2020/61), Taif University, Taif, Saudi Arabia.

Institutional Review Board Statement: Not applicable.

Informed Consent Statement: Not applicable.

Acknowledgments: The authors would like to acknowledge the financial support received from Taif University Researchers Supporting Project Number (TURSP-2020/61), Taif University, Taif, Saudi Arabia.

Conflicts of Interest: The authors declare no conflict of interest.

Appendix A. Consideration of Composite Loads

In this paper, the loads have been considered as constant power models, which is the approach frequently taken in similar papers. However, constant current, constant impedance, or a polynomial combination of the three models can be easily considered as follows:

$$P_i^{sch}(V_i) = P_{G_i} - P_{D_i}(V_i) = P_{G_i} - P_{0_i}\left[a_{0_i}\left(\frac{V_i}{V_{0_i}}\right)^2 + a_{1_i}\left(\frac{V_i}{V_{0_i}}\right) + a_{2_i}\right] \quad \text{(A1)}$$

$$Q_i^{sch}(V_i) = Q_{G_i} - Q_{D_i}(V_i) = Q_{G_i} - Q_{0_i}\left[b_{0_i}\left(\frac{V_i}{V_{0_i}}\right)^2 + b_{1_i}\left(\frac{V_i}{V_{0_i}}\right) + b_{2_i}\right] \quad \text{(A2)}$$

where P_{G_i} and Q_{G_i} are the active and reactive power generation at bus i, respectively. P_{D_i} and Q_{D_i} are the active and reactive power demand at bus i, respectively, P_{0_i} and Q_{0_i} are the active and reactive power consumption at rated voltage V_{0_i} at bus i, respectively. Parameters (a_{0_i}, b_{0_i}), (a_{1_i}, b_{1_i}), and (a_{2_i}, b_2) are used to model the constant impedance, constant current, and constant power loads, respectively.

Models (A1) and (A2) can be easily incorporated in the proposed methods by substituting these expressions in the system of Equation (1).

References

1. Tinney, W.F.; Hart, C.E. Power Flow solution by Newton's method. *IEEE Trans. Power Appar. Syst.* **1967**, *11*, 1449–1460. [CrossRef]
2. Stott, B.; Alsac, O. Fast decoupled load flow. *IEEE Trans. Power Appar. Syst.* **1974**, *93*, 859–869. [CrossRef]
3. van Amerongen, R.A.M. A General-Purpose Version of the Fast Decoupled Load Flow. *IEEE Trans. Power Syst.* **1989**, *4*, 760–770. [CrossRef]
4. Tortelli, O.L.; Lourenço, E.M.; Garcia, A.V.; Pal, B.C. Fast Decoupled Power Flow to Emerging Distribution Systems via Complex pu Normalization. *IEEE Trans. Power Syst.* **2015**, *30*, 1351–1358. [CrossRef]
5. da Costa, V.M.; Martins, N.; Pereira, J.L.R. Developments in the Newton Raphson power flow formulation based on current injections. *IEEE Trans. Power Syst.* **1999**, *14*, 1320–1326. [CrossRef]
6. Garcia, P.A.N.; Pereira, J.L.R.; Carneiro, S.; Vinagre, M.P.; Gomes, F.V. Improvements in the Representation of PV Buses on Three-Phase Distribution Power Flow. *IEEE Trans. Power Deliv.* **2004**, *19*, 894–896. [CrossRef]
7. Saleh, S.A. The Formulation of a Power Flow Using d-q Reference Frame Components—Part I: Balanced 3ϕ Systems. *IEEE Trans. Ind. Appl.* **2016**, *52*, 3682–3693. [CrossRef]
8. Saleh, S. A The Formulation of a Power Flow Using d-q Reference Frame Components—Part II: Unbalanced 3ϕ Systems. *IEEE Trans. Ind. Appl.* **2018**, *54*, 1092–1107. [CrossRef]
9. Pires, R.; Mili, L.; Chagas, G. Robust complex-valued Levenberg-Marquardt algorithm as applied to power flow analysis. *Int. J. Electr. Power Energy Syst.* **2019**, *113*, 383–392. [CrossRef]
10. Iwamoto, S.; Tamura, Y. A Load Flow Calculation Method for Ill-Conditioned Power Systems. *IEEE Trans. Power App. Syst.* **1981**, *100*, 1736–1743. [CrossRef]
11. Tripathy, S.C.; Prasad, G.D.; Malik, O.P.; Hope, G.S. Load-Flow Solutions for Ill-Conditioned Power Systems by a Newton-Like Method. *IEEE Trans. Power App. Syst.* **1982**, *10*, 3648–3657. [CrossRef]
12. Braz, L.M.C.; Castro, C.A.; Murati, C.A.F. A critical evaluation of step size optimization based load flow methods. *IEEE Trans. Power Syst.* **2000**, *15*, 202–207. [CrossRef]
13. Milano, F. Continuous Newton's Method for Power Flow Analysis. *IEEE Trans. Power Syst.* **2009**, *24*, 50–57. [CrossRef]
14. Tostado-Véliz, M.; Kamel, S.; Jurado, F. A robust Power Flow Algorithm Based on Bulirsch-Stoer Method. *IEEE Trans. Power Syst.* **2019**, *34*, 3081–3089. [CrossRef]
15. Tostado-Véliz, M.; Kamel, S.; Jurado, F. Robust and efficient approach based on Richardson extrapolation for solving badly initialized/ill-conditioned power-flow problems. *IET Gener. Transm. Distrib.* **2019**, *13*, 3524–3533. [CrossRef]
16. Tostado-Véliz, M.; Kamel, S.; Jurado, F. A powerful power-flow method based on Composite Newton-Cotes formula for ill-conditioned power systems. *Int. J. Electr. Power Energy Syst.* **2020**, *106*, 105558. [CrossRef]

17. Derakhshandeh, S.Y.; Pourbagher, R. Application of high-order Levenberg–Marquardt method for solving the power flow problem in the ill-conditioned systems. *IET Gener. Transm. Distrib.* **2016**, *10*, 3017–3022. [CrossRef]
18. Derakhshandeh, S.Y.; Pourbagher, R. A powerful method for solving the power flow problem in the ill-conditioned systems. *Int. J. Electr. Power Energy Syst.* **2018**, *94*, 88–96.
19. Tostado-Véliz, M.; Kamel, S.; Jurado, F. An effective load-flow approach based on Gauss-Newton formulation. *Int. J. Electr. Power Energy Syst.* **2019**, *113*, 573–581. [CrossRef]
20. Derakhshandeh, S.Y.; Pourbagher, R. Application of high-order Newton-like methods to solve power flow equations. *IET Gener. Transm. Distrib.* **2016**, *10*, 1853–1859. [CrossRef]
21. Tostado-Véliz, M.; Kamel, S.; Jurado, F. Developed Newton-Raphson based Predictor-Corrector load flow approach with high convergence rate. *Int. J. Electr. Power Energ Syst.* **2019**, *105*, 785–792. [CrossRef]
22. Ajjarapu, V.; Christy, C. The continuation power flow: A tool for steady state voltage stability analysis. *IEEE Trans. Power Syst.* **1992**, *7*, 416–423. [CrossRef]
23. Dukpa, A.; Venkatesh, B.; El-Hawary, M. Application of continuation power flow method in radial distribution systems. *Electr. Power Syst. Res.* **2009**, *79*, 1503–1510. [CrossRef]
24. Ju, Y.; Wu, W.; Zhang, B.; Sun, H. Loop-analysis-based continuation power flow algorithm for distribution networks. *IET Gener. Transm. Distrib.* **2014**, *8*, 1284–1292. [CrossRef]
25. Yang, X.; Zhou, X. Application of asymptotic numerical method with homotopy techniques to power flow problems. *Int. J. Electr. Power Energy Syst.* **2014**, *57*, 375–383. [CrossRef]
26. Tostado, M.; Kamel, S.; Jurado, F. Several robust and efficient load flow techniques based on combined approach for ill-conditioned power systems. *Int. J. Electr. Power Energy Syst.* **2019**, *110*, 349–356. [CrossRef]
27. Trias, A.; Marin, J.L. The Holomorphic Embedding Loadflow Method for DC Power Systems and Nonlinear DC Circuits. *IEEE Trans. Circuits Syst. I Regul. Pap.* **2016**, *63*, 322–333. [CrossRef]
28. Rao, S.; Feng, Y.; Tylavsky, D.J.; Subramanian, M.K. The Holomorphic Embedding Method Applied to the Power-Flow Problem. *IEEE Trans. Power Syst.* **2016**, *31*, 3816–3828. [CrossRef]
29. Chiang, H.-D.; Wang, T.; Sheng, H. A Novel Fast and Flexible Holomorphic Embedding Power Flow Method. *IEEE Trans. Power Syst.* **2017**, *33*, 2551–2562. [CrossRef]
30. Argawal, R.P.; O'Regan, D.; Sahu, D.R. Iterative Construction of Fixed Points of Nearly Asymptotically Nonexpansive Mappings. *J. Nonlinear Convex Anal.* **2007**, *8*, 61–79.
31. Tostado-Véliz, M.; Kamel, S.; Jurado, F. Powerful Power Flow Approach Based on the S-iteration Process. *IEEE Trans. Power Syst.* **2020**, *35*, 4148–4158. [CrossRef]
32. Sahu, D.R.; Singh, K.K.; Singh, V.K. Some Newton-like methods with sharper error estimates for solving operator equations in Banach spaces. *Fixed Point Theory Appl.* **2012**, *2012*, 78. [CrossRef]
33. Karakaya, V.; Dogan, K.; Atalan, Y.; Bouzara, N. The local and semilocal convergence analysis of new Newton-like iteration methods. *Turkish. J. Math.* **2018**, *42*, 735–751.
34. Saadat, H. *Power System Analysis*, 3rd ed.; PSA Publishing: Toronto, ON, Canada, 2011.
35. Cordero, A.; Hueso, J.L.; Martínez, E.; Torregrosa, J.R. A modified Newton-Jarratt's composition. *Numer. Algor.* **2010**, *55*, 87–99. [CrossRef]
36. McDougall, T.J.; Wotherspoon, S.J. A simple modification of Newton's method to achieve convergence of order $1 + \sqrt{2}$. *Appl. Math. Lett.* **2014**, *29*, 20–25. [CrossRef]
37. Lotfi, T.; Bakhtiari, P.; Cordero, A.; Mahdiani, K.; Torregrosa, J.R. Some new efficient multipoint iterative methods for solving nonlinear systems of equations. *Int. J. Comput. Math.* **2015**, *92*, 1921–1934. [CrossRef]
38. Zimmerman, R.D.; Murillo-Sanchez, C.E.; Thomas, R.J. MATPOWER: Steady-State Operations, Planning, and Analysis Tools for Power Systems Research and Education. *IEEE Trans. Power Syst.* **2011**, *26*, 12–19. [CrossRef]
39. Birchfield, A.B.; Xu, T.; Gegner, K.M.; Shetye, K.; Overbye, T. Grid Structural Characteristics as Validation Criteria for Synthetic Networks. *IEEE Trans. Power Syst.* **2017**, *32*, 3258–3265. [CrossRef]
40. Josz, C.; Fliscounakis, S.; Maeght, J.; Panciatici, P. AC Power Flow Data in Matpower and QCQP Format: iTesla, RTE Snapshots, and PEGASE. *arXiv* **2016**, arXiv:1603.01533. Available online: http://arxiv.org/abs/1603.01533 (accessed on 6 November 2021).
41. Fliscounakis, S.; Panciatici, P.; Capitanescu, F.; Wehenkel, L. Contingency Ranking with Respect to Overloads in Very Large Power Systems Taking Into Account Uncertainty, Preventive, and Corrective Actions. *IEEE Trans. Power Syst.* **2013**, *28*, 4909–4917. [CrossRef]
42. Tostado-Véliz, M.; Hasanien, H.M.; Turky, R.A.; Alkuhayli, A.; Kamel, S.; Jurado, F. Mann-Iteration Process for Power Flow Calculation of Large-Scale Ill-Conditioned Systems: Theoretical Analysis and Numerical Results. *IEEE Access* **2021**, *9*, 132255–132266. [CrossRef]
43. Ezquerro, J.; Hernández, M. An optimization of Chebyshev's method. *J. Complex.* **2009**, *25*, 343–361. [CrossRef]
44. Tostado-Véliz, M.; Alharbi, T.; Alrumayh, O.; Kamel, S.; Jurado, F. A Novel Power Flow Solution Paradigm for Well and Ill-Conditioned Cases. *IEEE Access* **2021**, *9*, 112425–112438. [CrossRef]
45. Tostado-Veliz, M.; Kamel, S.; Jurado, F. Two Efficient and Reliable Power-Flow Methods with Seventh Order of Convergence. *IEEE Syst. J.* **2020**, *15*, 1026–1035. [CrossRef]

Article

Parametric Analysis for Performance Optimization of Line-Start Synchronous Motor with Interior Asymmetric Permanent Magnet Array Rotor Topology

Vasilija Sarac [1,*], Dragan Minovski [1] and Peter Janiga [2]

1 Faculty of Electrical Engineering, University Goce Delcev, 2000 Stip, North Macedonia; dragan.minovski@ugd.edu.mk
2 Faculty of Electrical Engineering and Information Technology, Slovak University of Technology in Bratislava, 81219 Bratislava, Slovakia; peter.janiga@stuba.sk
* Correspondence: vasilija.sarac@ugd.edu.mk

Abstract: Line-start synchronous motors have attracted researchers' interest as suitable replacements of asynchronous motors due to their high efficiency, which has been provoked by strict regulations regarding applicable efficiency classes of motors in the EU market. The research becomes even more challenging as it takes into consideration the diverse rotor topologies with different magnet locations for this type of motor. The rotor configuration with an interior asymmetric permanent magnet (PM) array rotor was chosen for analysis and optimization in this paper as this specific configuration is particularly challenging in terms of placing the magnets with adequate dimensions into the existing rotor of the asynchronous motor with a squirrel cage winding, in order simultaneously to obtain good operational characteristics such as high efficiency and power factor, good overloading capability and low material consumption. Therefore, an optometric analysis is performed in order to find the best configuration of the air gap length, magnet thickness, magnet width and number of conductors per slot, along with modifications of the rotor slot. The motor outer dimensions remained unchanged compared with the starting model of the line-start motor derived from the asynchronous motor, which is a product of the company Končar. The optimized model obtained higher efficiency, power factor and overloading capability than the starting model, along with good starting and synchronization capabilities.

Keywords: line-start synchronous motor; efficiency factor; power factor; optometric analysis; transient models

Citation: Sarac, V.; Minovski, D.; Janiga, P. Parametric Analysis for Performance Optimization of Line-Start Synchronous Motor with Interior Asymmetric Permanent Magnet Array Rotor Topology. *Electronics* **2022**, *11*, 531. https://doi.org/10.3390/electronics11040531

Academic Editor: Ahmed F. Zobaa

Received: 26 January 2022
Accepted: 5 February 2022
Published: 10 February 2022

1. Introduction

Line-start synchronous motors have gained popularity as an alternative to asynchronous squirrel cage motors, especially in constant speed applications, due to the strict regulations that have been imposed worldwide regarding the efficiency classes of motors that can be used. The asynchronous squirrel cage motors can achieve the IE3 or IE4 efficiency class (in general, efficiency above 89%) with numerous modifications which increase the motor dimensions, material consumption or even imply usage of more expensive materials such as copper bars in the squirrel cage winding or steel laminations with low losses. From 1 July 2021, low-voltage motors up to 1000 kW must meet at least efficiency class IE3 according to a new EU Directive. In a second step, from mid-2023, efficiency class IE4 will become mandatory for the 75–200 kW performance range. Some manufacturers of the motors have answered these challenges and have offered to the market three-phase asynchronous induction motors of the IE4 class [1,2]. Researchers have also analyzed various modifications of motor slots such as adding magnet wedges in induction motors with semi-closed slots in order to reduce copper and core losses and increase motor efficiency [3]. The IE4 efficiency class can be more easily achieved with the line-start synchronous motor

(LSSM) as there is no induced current in the rotor winding due to the synchronous speed of rotation, so the rotor copper losses are nullified [4]. The high power factor at LSSM allows smaller line current and lower copper losses in the stator winding, that in turn increase the efficiency factor; therefore, this type of motor can easily achieve efficiency class IE4 (in general, efficiency above 89%). The combination of the squirrel cage winding and the magnets in the rotor allows for direct starting of the motor with voltage from the mains without the need of the voltage inverters, which are typically needed for starting synchronous motors without squirrel cage winding, as well as synchronization of the motor, provided by the magnets that pull the motor into synchronism. Yet, the proper design of cage winding and the magnets is essential, as the magnets produce the breaking torque that lowers the motor's starting torque and prolongates the motor starting; however, their improper design results in failure of motor synchronization [5–9]. Not only are the magnets responsible for motor operating regimes, starting and steady-state; the stator winding turns also affect the winding resistance, current and the power factor [10]. Another aspect of motor operation is the material of the squirrel cage winding, which is usually aluminum or copper that affects the motor starting and the temperature distribution [11]. The temperature distribution also affects the partial demagnetization of the magnets and the operation of the motor, and one such example is analyzed in [12]. Other authors propose innovative solutions regarding rotor design that include two different types of rotor slots or that focus on the optimization of the rotor slot that allows the best operating characteristics of the motor [13,14]. Various optimization techniques have been implemented in the optimization of the magnet thickness, magnet width or the rotor slots at line-start synchronous motors with hybrid magnets (combination of two types of magnet materials) or the line-start synchronous motor with a configuration of magnets with radial flux distribution [15–17]. A very small number of works can be found regarding optimization of LSSM with asymmetric permanent magnet array topology. Optimization of the flux barriers of LSSM with asymmetric permanent magnet array topology, which decreases the flux leakage, is found to be a good optimization approach in efficiency optimization, together with the optimization of the dimensions of the rotor slot in [18]. This paper presents a two-step design modification of LSSM with asymmetric permanent magnet array topology. Authors' previous research has shown that there are some differences regarding motor operating characteristics and material consumption in correlation with the specific rotor topology of LSSM [19]. The starting point of the analysis is a three-phase squirrel cage motor type 5AZ 100LA-4, which is a product of the company Rade Končar. The design of the asynchronous motor was modified with a rotor with an asymmetric permanent magnet array topology, thus obtaining the starting model of LSSM (BM). The main constrain of the motor design is the new derived LSSM having the same output power as the asynchronous motor of 2.2 kW. The laminations of the stator and the rotor were obtained from Končar and they remain unchanged in the process of modification of the asynchronous motor into LSSM. The BM, due to the limited space for magnet placement, imposed by the dimensions and shape of rotor slots (Končar design), has a relatively low consumption of permanent magnet material, but poor overloading capability, although the efficiency and the power factor are high. Therefore, as the first step in the design modification was to modify the rotor slots in order to provide more space for magnets and flux barriers. Apart from rotor slot modification and magnet dimensions, no other modifications were made in the design of this second model (M1). The model M1 has good efficiency and an improved power factor and overloading capability but has the relatively high consumption of a permanent magnet material. Therefore, the second step in the design modification was to run the optometric analysis of the M1 model where the outer rotor diameter, magnet thickness and width, along with number of conductors per stator slot, are varied simultaneously within predefined limits and the overload capability, efficiency, power factor and magnet consumption are followed in each combination (iteration of model solving) of those four varied parameters. A total of 25,257 combinations were solved, resulting in model M2, which was found to have the highest efficiency factor, and a good power factor and overloading capability, along with low consumption of permanent magnet material. Optometric analysis is a software module within Ansys Electronics Desktop software; more precisely, it is included in the RMxprt module of the Ansys software and allows arbitrary machine parameters to be varied within defined boundaries while the arbitrary machine characteristics such as efficiency, power factor or overloading capability, depending on the designer's point of interest, are calculated for each combination of the varied parameters. In this way, the designer can choose the best combination of the motor parameters (for example, outer rotor diameter, number of conductors per slot, magnet thickness and magnet length) that produce the best performance in the machine, for example, highest efficiency, power factor or overloading capability. All motor models are analyzed for the flux density distribution by FEM. The transient characteristics of all motor models were derived, allowing analysis of motor operation at start-up and synchronization. The redesign of the rotor of the asynchronous motor for obtaining LSSM needs careful evaluation and analysis, especially when various rotor topologies are available in order to obtain optimal results regarding motor operation and material consumption.

2. Computer Models for Steady-State and Transient Characteristics

One part of the redesign of the three-phase asynchronous squirrel cage motor into LSSM is to place the magnets inside the rotor, which along with squirrel cage winding, allow starting and synchronization of the motor. The starting point in the analysis was the three-phase squirrel cage motor, a product of Rade Končar, type 5AZ100LA-4, 2.2 kW, 1410 rpm, 5 A, power factor of 0.83, efficiency of 79%. The topology with asymmetric permanent magnet array was chosen for the rotor redesign as the authors' previous research showed that this topology regarding the analyzed type of the asynchronous motor has some drawbacks, including low overloading capability and relatively low power factor [19]. Therefore, it was a challenging task to improve the overloading capability, power factor and efficiency with minimum consumption of permanent magnet material while keeping the same power output of 2.2 kW, as it is in the asynchronous motor. The LSSM with the asymmetric permanent magnet array topology is presented in Figure 1a.

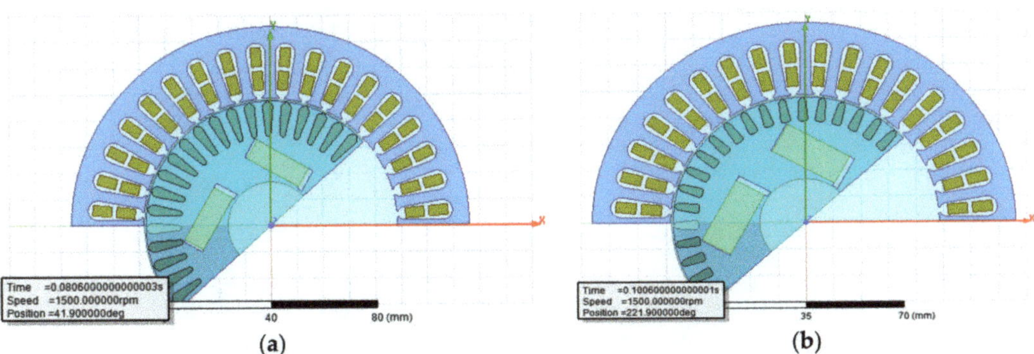

Figure 1. Cross section of line-start synchronous motor with interior asymmetric permanent magnet array rotor topology (**a**) Model BM (**b**) Model M1.

Firstly, the computer model of the asynchronous motor for calculating motor parameters and steady-state characteristics was modeled. This model will be referred to as AM. Since all the further computer models of the line-start synchronous motor will be derived from this model (AM), it was necessary to verify its accuracy by comparing data obtained from AM with the catalogue data from the producer of the motor [20]. This comparison and the obtained results are presented in Table 1. From the results presented in Table 1, it can be concluded that the AM model is sufficiently accurate and it can be further modified into line-start synchronous motor with asymmetric permanent magnet array rotor topology. The BM model of line-start synchronous motor is derived from three-phase asynchronous squirrel cage motor 5AZ 100LA-4, a product of Končar, without any alteration of the stator dimensions or the geometry of the stator and rotor slots [20]. The BM model is derived for the same output power of 2.2 kW as it is in the asynchronous motor. The model is designed to obtain the highest possible efficiency and output power with minimum consumption of permanent magnet material. Therefore, in the asynchronous motor (AM), whose data are presented in Table 1, the rotor is modified by decreasing its diameter, i.e., the air gap length is increased from 0.3 mm to 1 mm along with adding the flux barriers and permanent magnets in asymmetric array topology. The increase in air gap was due to the modification of asynchronous motor in line-start synchronous motor in order to maintain the good overloading capability of the line-start synchronous motor. The AM has an overload capability (maximum torque versus nominal torque) of 2.6. The dimensions of the magnets, height and thickness, were calculated for obtaining the highest efficiency and power factor along with a good overloading capability of the motor. This first model in the analysis of the line-start synchronous motor derived from the asynchronous motor, without the changes of the dimensions of the stator, slots of the rotor and the stator, is referred to as the BM model. This BM model will be the starting point with which the modified and optimized models will be compared.

Table 1. Data of asynchronous motor.

Parameter	AM	Producer
Nominal power (kW)	2.2	2.2
Number of poles 2p (/)	4	4
Nominal voltage Δ/Y (V)	220/380	220/380
Nominal current Δ/Y (A)	9/5.2	8.3/4.8
Power factor (/)	0.8	0.83
Nominal speed (rpm)	1353	1410
Conductors per slot CPS [/]	115	/
Stator winding resistance at 20 °C (Ω)	2.75	/
Stator copper losses (W)	277	/
Rotor copper losses (W)	240	/
Iron core losses (W)	28	/
Frictional and windage losses (W)	22	/
Stray losses (W)	39.6	/
Efficiency (%)	78.4	79.7
Rated torque (Nm)	15.5	14.9
Locked-Rotor Torque Ratio (/)	2.5	2.2
Locked-Rotor Current ratio (A)	4.2	5.2
Radial air gap length (mm)	0.3	/
Break-Down Torque Ratio (/)	2.65	2.7

The analytical calculations for this model were performed in Ansys software together with the calculation of steady-state characteristics. Therefore, it was necessary to input the exact dimensions of the cross-section of the motor along with the characteristics of all materials applicable in the motor design. The results obtained from the BM regarding parameters and operating characteristics showed that although BM has low consumption of permanent magnet material, it has little overloading capability. The consumption of permanent magnet material is limited due to available space for the placement of magnets in the rotor and this affects the overloading capability of the motor. This is due to the geometry of the rotor slot (Figure 1a) which was taken over from the asynchronous motor. Therefore, the first step in improving the model of the line-start synchronous motor with asymmetric permanent magnet array rotor topology was to modify the rotor slots, keeping almost the same cross-section of the slot with a modification in its geometry that allows more space in the rotor for the magnets to be placed. This second model will be referred to as M1 model. The modification of the slot in M1 in comparison to BM is presented in Figure 1b. In addition to rotor slot modification in M1, all other dimensions of the motor, features of the both windings, and material properties remain unchanged. From the results obtained from model M1, it was observed that this model had significantly larger consumption of permanent magnet material than B1, although the overloading capability was improved. The second step in model improvement was to run the optometric analysis of the model M1 where four parameters are chosen to be varied simultaneously within predefined ranges: the outer rotor diameter (ORD), magnet thickness (MT), magnet width (MW) and number of conductors per stator slot (CPS). The motor variables and their ranges of variation are presented in Table 2.

Table 2. Motor parameters and their ranges of variation at M1.

Parameter	Variation Range
Number of conductors per slot (/)	65 ÷ 105
Outer rotor diameter (mm)	95 ÷ 96
Magnet thickness (mm)	22.5 ÷ 25.5
Magnet width (mm)	8 ÷ 11.5

The ranges of variation of magnet geometry were defined on the base of the available space in the rotor. The computer program calculates the slot fill factor for the stator slots. The program is set to the maximum slot fill factor of 75%. When the limit is reached, the program adjusts the wire diameter in order not to succeed the limit of the slot fill factor. The rotor's outer diameter is varied, taking into consideration the inner stator diameter and the air gap length of the asynchronous motor. The length of the air gap of the asynchronous motor is 0.3 mm. The overloading capability was one of the issues that needed to be improved in the optimized model of line start synchronous motor. The larger air gap contributes to the increased overloading capability while simultaneously worsening the efficiency factor and power factor. Another design aspect is the dimension of the magnets. The increased magnet thickness has a positive effect on the increase in the efficiency, power factor and the overloading capability of the motor; however, it decreases the starting torque. The increase in the magnet width decreases the efficiency but increases the overloading capability and the power of the motor. From the above, it is obvious that various selected parameters have a contradictory effect on motor operating characteristics and there is no straightforward solution which combines the four above-mentioned varied parameters and produces the best operating characteristics of the motor at steady-state operation as well as at transient regimes. Therefore, by optometric analysis, each combination of motor variables (25,257 combinations) is implemented in motor analytical model, modeled in Ansys software and, by following the output results with respect to motor operating characteristics, the most favorable analytical motor model can be selected for further analysis by the aid of numerical and dynamic models. A total of 25,257 model combinations with the varied parameters were solved. Among all 25,257 models, the three most favorable solutions were chosen in terms of the efficiency, the power factor, the overloading capability and the consumption of permanent magnet material. These models will be referred to as models M2, M3 and M4. In terms of the highest efficiency factor and the smallest permanent magnet material consumption, model M2 has the best results; therefore, this model is further analyzed with numerical methods and applied into the simulation circuits of the dynamic models. The basic criteria for choosing the models M2 to M4 (obtained by optometric analysis) was to have an overloading capability above 2.2, efficiency above 95.9 and power factor above 0.9. Model M1 is derived from model BM without any optometric analysis, only by redesigning the rotor slots, in order for more magnet material to be placed, so as to obtain a larger overloading capability than the BM model. This was achieved, either because the M1 model has a maximum output power of 4326 W compared to 3572 W of the BM model, or because the M1 model has an increased overloading capability of 1.9 compared to the BM model which has 1.6. Another aspect of motor design is the permanent demagnetization of magnets due to reverse fields exceeding the value of H_d, a point at which the magnetic vector polarization vector M collapses. The corresponding value of flux density is B_d. The demagnetization of magnets has been checked according to [21]:

$$I_{dgmrm} = \frac{p\pi}{6\mu_0(K_{w1}N_c)}(B_r h_m - B_d(g + h_m)) \tag{1}$$

where I_{dgmrm} is the maximum permitted value of steady-state stator current for normal steady-state operation before demagnetization (A). p is the number of stator poles, K_{w1} is the winding factor, N_c number of turns per phase of stator winding, h_m is the magnet thickness in radial direction in meters, g is the air gap length in meters and B_r is the residual flux density in Tesla at the operating temperature of the magnet. In all motor models, the SmCo28 magnets are used with remanent flux density of 1.07 T and coercitivity of 820,000 A/m. The analysis of demagnetization of the magnets is especially important during transient regimes, i.e., at motor starting and synchronization. At asynchronous starting, currents with a maximum value up to several times greater than the amplitude of the rated motor current can flow in the stator winding. Supply voltage, the magnetic flux generated by magnets, the moment of inertia of rotating masses and load torque affect the start-up course and the amplitude of the inrush current. The impact of the magnetomotive force caused by the armature interaction related to the amplitude of the stator currents may cause partial demagnetization of the permanent magnets located in the motor [12]. Due to irreversible demagnetization of the magnets, the main magnetic flux is irreversibly reduced and consequently so is the motor torque. The demagnetization of the magnets at motor starting and in the vicinity of synchronization speed is analyzed by FEM. The flux density at magnets at various speeds during acceleration of model M2 is presented in Figure 2 for a load torque of 14 Nm and a moment of inertia of 0.37 kgm^2. The magnitude of the magnetic field for the same operating regimes from Figure 2 is presented in Figure 3. From the presented results in Figures 2 and 3 and for the type of magnets used, the partial demagnetization could occur in tiny areas of magnet edges in the vicinity of synchronous speed. In other analyzed operating points during motor acceleration, the demagnetization of magnets should not occur.

The numerical model allows magnetic flux density distribution to be calculated in the motor cross-section by the aid of Finite Elements Method (FEM) thus allowing parts of the magnetic core with high flux density to be detected [22–24]. Another aspect of analysis of the models is the transient characteristics where the motor behavior in transient regimes such as start-up can be analyzed [25,26]. The M2 model is implemented in the dynamic model in order to obtain transient characteristics of speed, current and torque at motor acceleration and at steady-state operation (operation with synchronous speed). The last part of the analysis is necessary due to the specific construction of line-start synchronous motor. Namely, the squirrel cage winding contributes to the motor starting directly with the voltage from the three phase supply, while permanent magnets pull the motor into synchronism. The magnets generate the breaking torque that can worsen the motor starting conditions. On the other hand, their improper design may result in the failure of motor synchronization. The dynamic models are designed for BM and M2 when the motor is accelerated with various loads and load inertia. The motor acceleration and synchronization is analyzed and adequate conclusions are derived. The motor dynamic model is presented in Figure 4. The dynamic model of the motor that allows calculation of transient characteristics is derived in Ansys Simplorer. The software has blocks that allow for modeling the symmetrical three-phase power supply. Additionally, it allows the model of the motor derived in RMxprt module of Ansys Electronic Desktop to be imported via a dynamic link. The motor output is linked to the load torque and inertia. The more detailed explanation how to derive drive design can be found in [27]. The computational time of the drive system with various loads and moments of inertia takes no more than couple of minutes depending of the set time for simulation and the set maximum and minimum time steps.

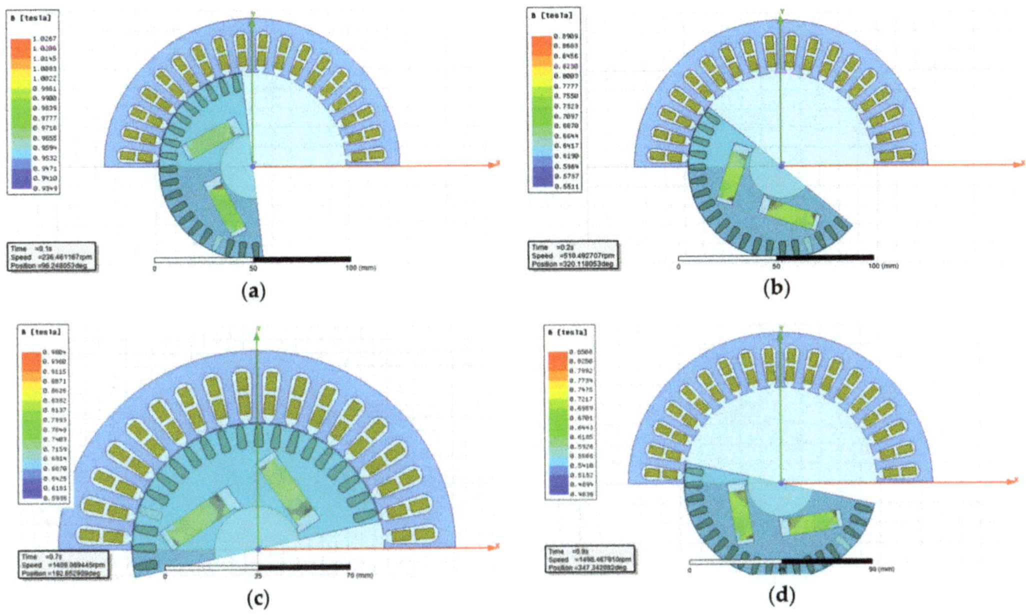

Figure 2. Flux density distribution at magnets of M2 at M_{load} = 14 Nm, J = 0.37 kgm^2 (**a**) 236 rpm (**b**) 510 rpm (**c**) 1409 rpm (**d**) 1498 rpm.

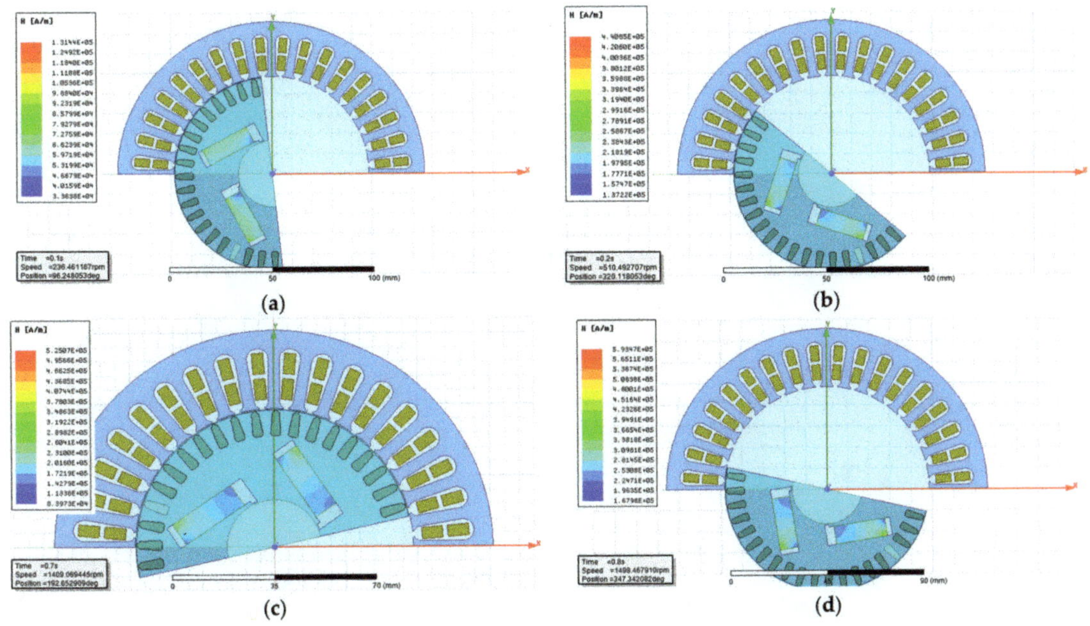

Figure 3. Magnetic field distribution at magnets of M2 at M_{load} = 14 Nm, J = 0.37 kgm^2 (**a**) 236 rpm (**b**) 510 rpm (**c**) 1409 rpm (**d**) 1498 rpm.

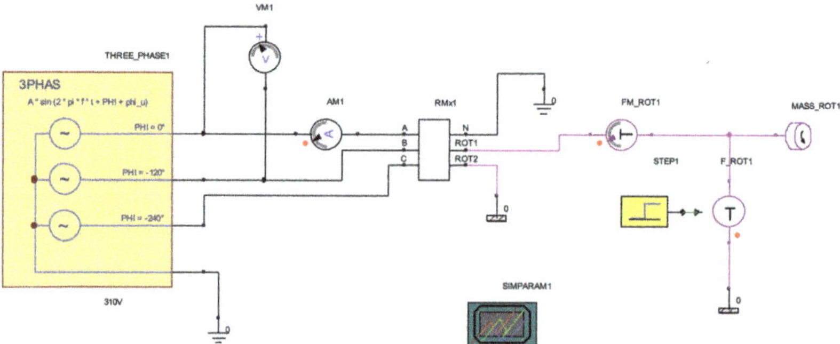

Figure 4. Dynamic model of line-start synchronous motor.

The basic equation behind the dynamic model of the motor in *d-q* reference frame is [28]:

$$u_{ds} = R_s i_{ds} + \frac{d\Psi_{ds}}{dt} - (1-s)\omega_s \psi_{qs} \tag{2}$$

$$u_{qs} = R_s i_{qs} + \frac{d\Psi_{qs}}{dt} + (1-s)\omega_s \psi_{ds} \tag{3}$$

where s is the slip defined as:

$$s = \frac{\omega_s - \omega_r}{\omega_s} \tag{4}$$

$$u_{dr} = R_{dr} i_{rd} + \frac{d\Psi_{dr}}{dt} = 0 \tag{5}$$

$$u_{qr} = R_{qr} i_{qd} + \frac{d\Psi_{qr}}{dt} = 0 \tag{6}$$

$$\Psi_{ds} = L_{ds} i_{ds} + L_{md} i_{dr} + \Psi_m \tag{7}$$

$$\Psi_{qs} = L_{qs} i_{qs} + L_{mq} i_{qr} \tag{8}$$

$$\Psi_{dr} = L_{dr} i_{dr} + L_{md} i_{ds} + \Psi_m \tag{9}$$

$$\Psi_{qr} = L_{qr} i_{dr} + L_{mq} i_{qs} \tag{10}$$

The coupling between the electrical system and the mechanical system is represented by the torque equation and the mechanical equation. The electromagnetic torque T_{el} developed by the motor can be expressed as:

$$T_{el} = \frac{p}{2}\frac{3}{2}(\Psi_{ds} i_{qs} - \Psi_{qs} i_{ds}) \tag{11}$$

The motor torque is balanced by the mechanical shaft torque T_{load} and the dynamic torque caused by the total inertia J [29].

$$T = T_{load} + J\left(\frac{2}{p}\right)\frac{d\omega_m}{dt} \tag{12}$$

Motor parameters are calculated in RMxprt module of Ansys Electronics Desktop software. This model of the motor from RMxprt module with all data and calculated parameters is linked, i.e., imported in the dynamic model which is modeled in the software module Ansys Simplorer. The motor parameters for the model M2 are presented in Table 3.

Table 3. Motor parameters of model M2.

Parameter	Value
D-axis reactive reactance X_{sad} (Ω)	18.9
Q-axis reactive reactance X_{saq} (Ω)	90.03
Q-axis reactance $X_1 + X_{aq}$ (Ω)	91.7
D-axis reactance $X_1 + X_{ad}$ (Ω)	20.6
Armature leakage reactance X_1 (Ω)	1.67
D-axis rotor resistance R_{dr} (Ω)	2.4
D-axis rotor leakage reactance X_{dr} (Ω)	0.83
Q-axis rotor resistance R_{qr} (Ω)	2.41
Q-axis rotor leakage reactance X_{qr} (Ω)	0.83
Stator winding phase resistance R_s (Ω)	1.15

The description of methodology for obtaining the numerous parameters of line-start synchronous motor can be found in [30]. Due to extent of the mathematical model, it is not presented here. Further details can be found in [30,31].

3. Results

3.1. Parameters, Optometric Analysis and Steady-State Characteristics

The computer models for BM, M1, M2, M3 and M4 that allow calculation of motor parameters and steady state characteristics are the first part of an analysis of all motor models. The motor parameters and operating characteristics are obtained as output data from these computer models for analytical calculation of the models. The obtained parameters and operating characteristics are presented in Table 4 for the rated load operating regime.

Table 4. Motor parameters and characteristics at rated load.

Parameter/Characteristics	BM	M1	M2	M3	M4
Air gap length (mm)	1	1	0.5	0.5	0.5
Outer rotor diameter (mm)	95	95	96	96	96
Magnet width (mm)	22.5	25.5	24.5	25.5	24.5
Conductors per slot (/)	97	97	73	78	73
Magnet thickness (mm)	8	11.5	8.5	9.5	11.5
Frictional and windage loss (W)	22	22	22	22	22
Iron core loss (W)	15.55	18.5	17	17.8	18
Copper loss (W)	95	86	51	54	48
Stator winding resistance at 20 °C (Ω)	1.7	1.7	0.95	1.1	0.95
Input power (W)	2333	2326	2290	2293	2288
Output power (W)	2200	2200	2200	2200	2200
Power factor (/)	0.9	0.95	0.91	0.94	0.93
Efficiency (%)	94.3	94.6	96.06	95.9	96.1
Current (A)	3.9	3.7	3.85	3.7	3.7
Rated torque (Nm)	14	14	14	14	14
Maximum output power (W)	3572	4326	4929	4873	6158
Starting torque (Nm)	61	64	112	98	111
Magnet weight (kg)	0.6	0.97	0.69	0.8	0.94
Torque angle (°)	79	71.5	78.5	77.2	74

Models M2, M3 and M4 are derived by taking into consideration the optometric analysis, i.e., the results obtained from it (the value of CPS, MT, MW and ORD for the adequate motor model). The M2 model has the biggest efficiency and overloading capability along with the smallest consumption of the permanent magnet material. This model is chosen for further comparison with the model BM. The data for the stator winding of model M2 are presented in Table 5. The data of stator, rotor and magnets geometry for model M2 are presented in Table 6. The presented data in Table 6 are supported by drawings of stator and rotor slot of model M2 presented in Figure 5.

Table 5. Data of stator, rotor winding and magnets of model M2.

Parameter	Value
Winding layers	2
Coil pitch	7
Number of parallel branches	2
Conductors per slot	73
Number of wires per conductor	3
Wire diameter (mm)	0.57

Table 6. Dimensions of stator and rotor lamination and magnets.

Parameter	Value
Stator outer diameter (mm)	152
Stator inner diameter (mm)	97
Rotor outer diameter (mm)	96
Rotor inner diameter (mm)	33
Motor length (mm)	100
Number of stator slots (/)	36
Number of rotor slots (/)	40
D1 (mm)	76
O1 (mm)	18
O2 (mm)	4
B1 (mm)	2
Rib (mm)	0.5
Magnet thickness (mm)	24.5
Magnet height (mm)	8.5
Stator slot	
Hs0 (mm)	0.5
Hs2 (mm)	14.7
Bs0 (mm)	2
Bs1 (mm)	5
Bs2 (mm)	7.6
Rotor slot	
Hs0 (mm)	0.5
Hs01 (mm)	0
Hs1 (mm)	0.5
Hs2 (mm)	7
Bs0 (mm)	1
Bs1 (mm)	2.5
Bs2 (mm)	3.5
Rs (mm)	1

The steady-state characteristic of the efficiency, power factor and torque for BM and M2 are presented in Figures 6–8, respectively. The steady-state characteristics should support the data, presented in Table 4, i.e., for the adequate torque angle, the rated values of efficiency, power factor and torque should be obtained.

The impact of each varied parameter (CPS, ORD, MT and MW) on efficiency, power factor, starting torque and maximum output power is analyzed by varying the parameter of interest within the prescribed limits while the rest three parameters are constant and equal to the values presented in Table 4 for the model M2. The analysis originates directly from the optometric analysis and allows for the impact of each parameter on motor operating characteristics to be evaluated separately. The impact of CPS on efficiency, power factor, starting torque and maximum output power is presented in Figure 9.

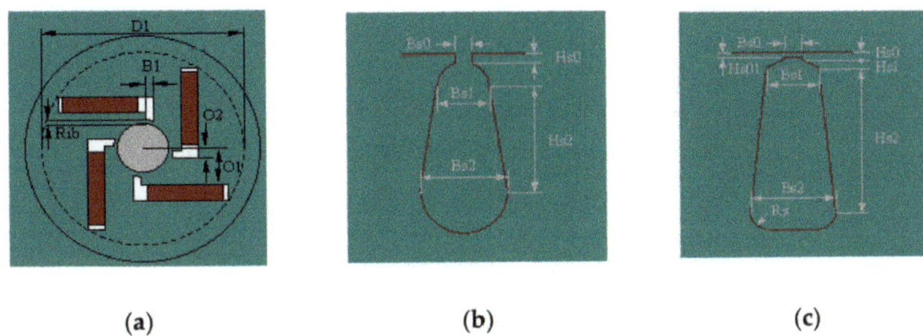

(a) (b) (c)

Figure 5. Dimensions (**a**) magnets (**b**) stator slot (**c**) rotor slot.

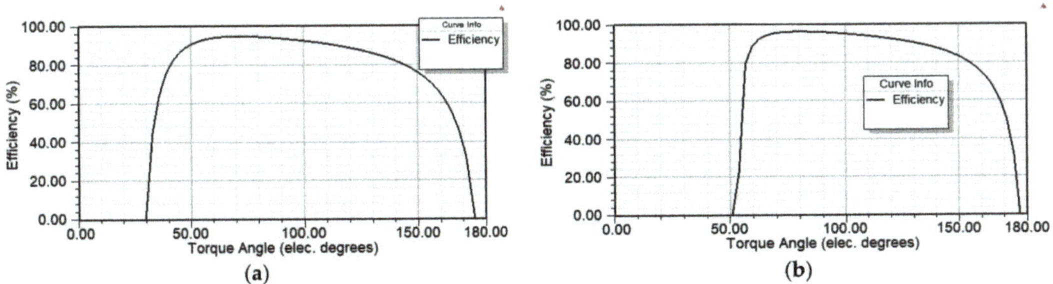

(a) (b)

Figure 6. Steady-state characteristics of efficiency (**a**) BM (**b**) M2.

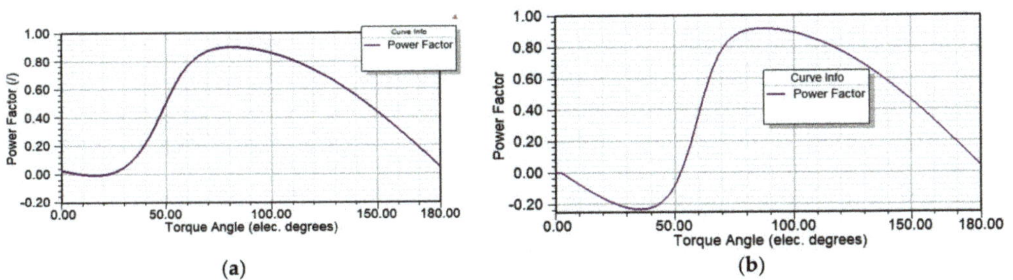

(a) (b)

Figure 7. Steady-state characteristics of power factor (**a**) BM (**b**) M2.

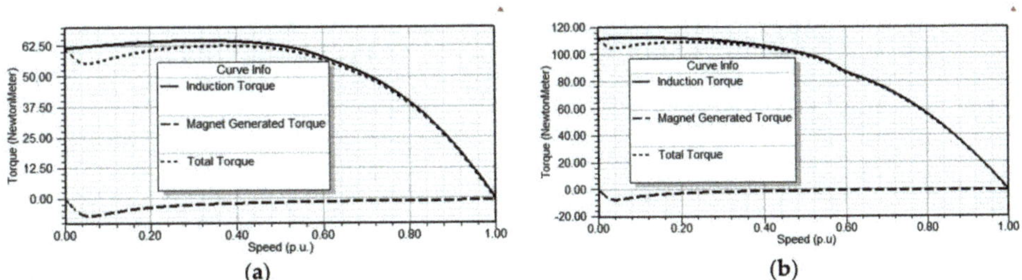

(a) (b)

Figure 8. Steady-state characteristics of torque (**a**) BM (**b**) M2.

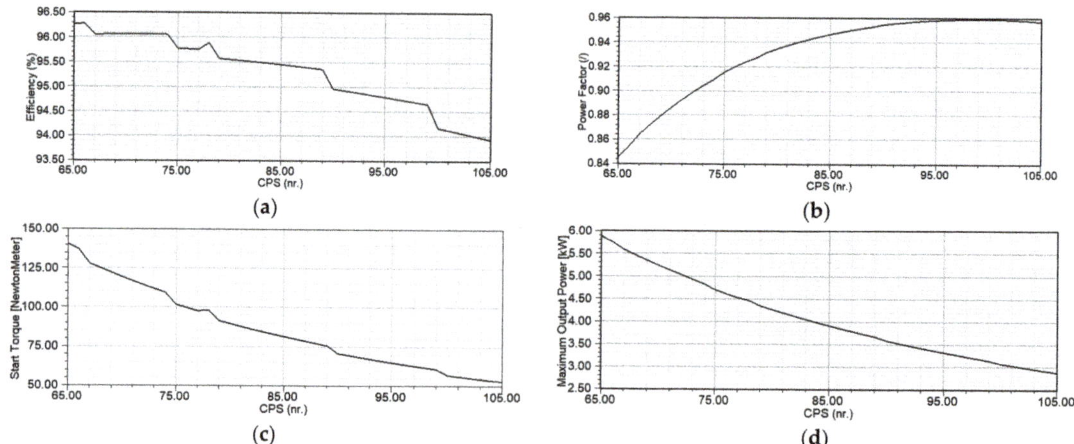

Figure 9. Impact of CPS on (**a**) efficiency (**b**) power factor (**c**) starting torque (**d**) maximum output power.

A similar analysis is performed for determining the impact of ORD, i.e., air gap length, on efficiency, power factor, starting torque and maximum output power. The obtained results are presented in Figure 10. The impact of MW on efficiency, power factor, starting torque and maximum output power of the motor is presented in Figure 11. The similar analysis for the impact of MT is presented in Figure 12.

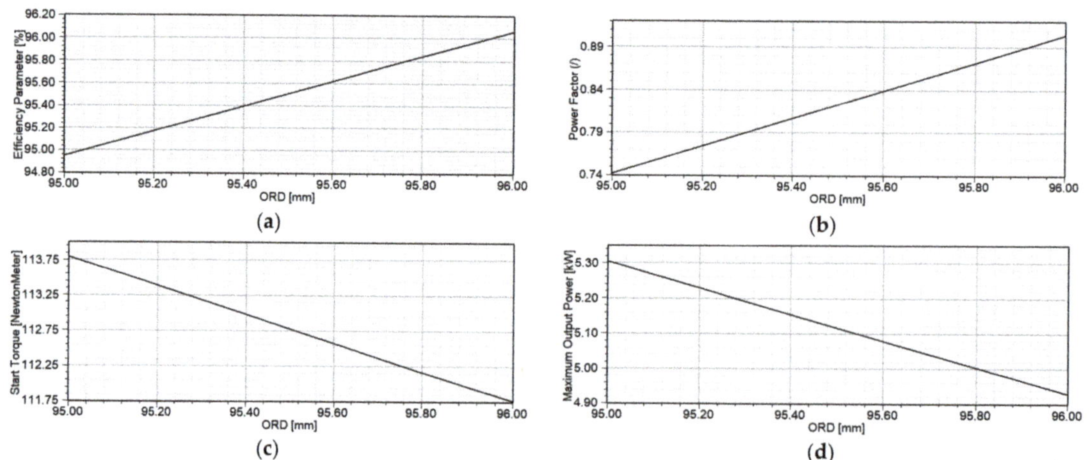

Figure 10. Impact of ORD on motor operating characteristics (**a**) efficiency (**b**) power factor (**c**) starting torque (**d**) maximum output power.

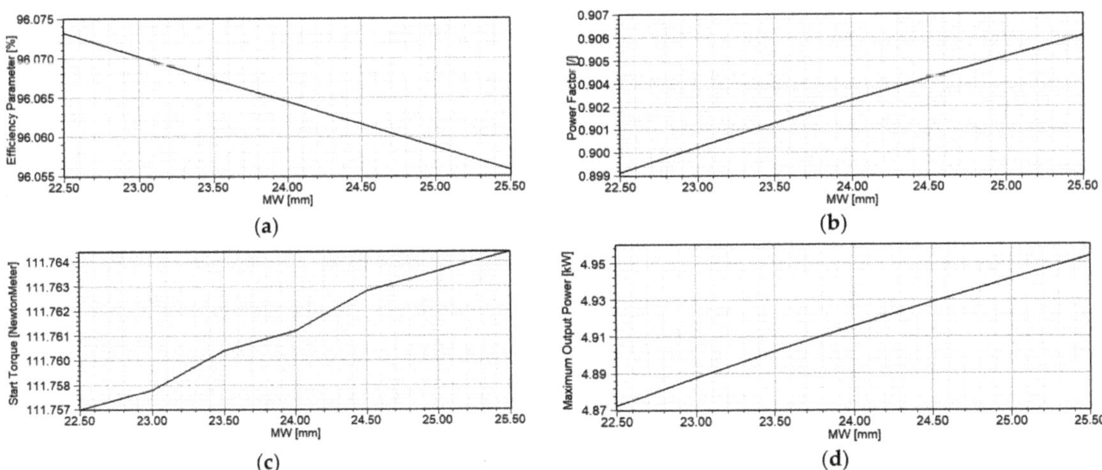

Figure 11. Impact of MW on motor operating characteristics (**a**) efficiency (**b**) power factor (**c**) starting torque (**d**) maximum output power.

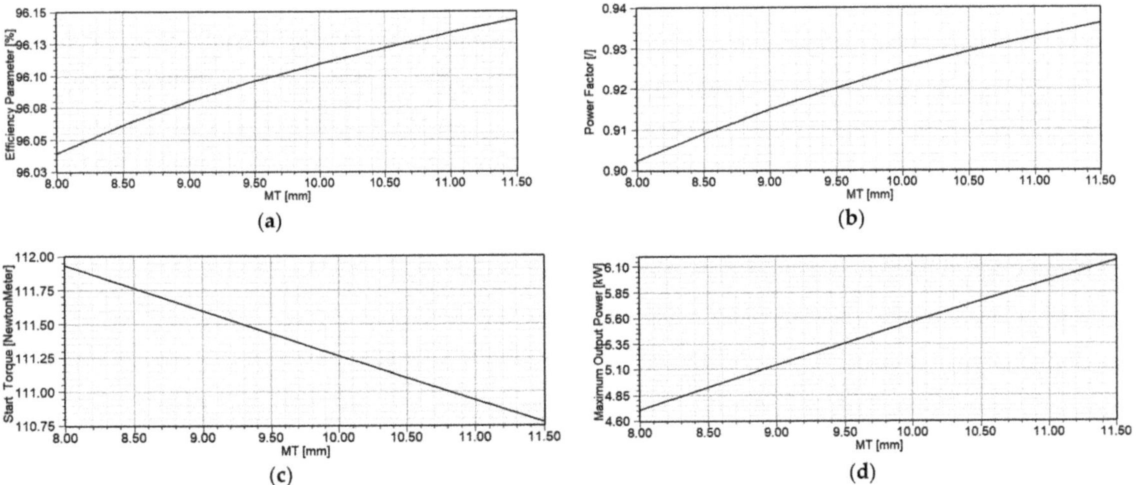

Figure 12. Impact of MT on motor operating characteristics (**a**) efficiency (**b**) power factor (**c**) starting torque (**d**) maximum output power.

3.2. FEM and Transient Models

The FEM models of the motors allows for the calculation of magnetic flux density distribution in the cross-section of motor models by solving the magnetic vector potential in the small areas of the mesh that are created in the cross-section of motors. The obtained results of magnetic flux density distribution in the motor cross-section are presented in Figure 13.

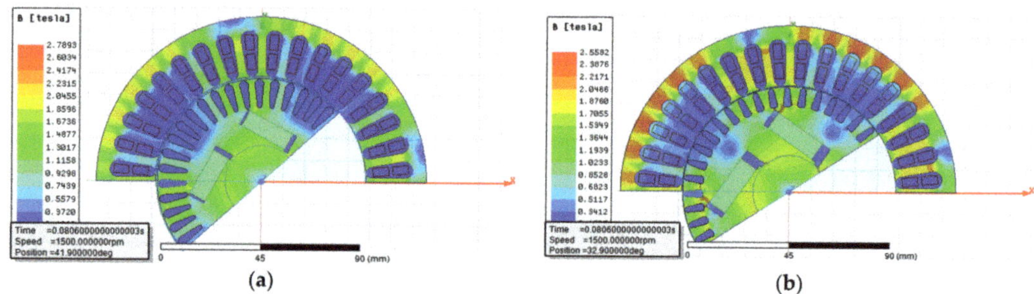

Figure 13. Flux density distribution (**a**) BM (**b**) M2.

The transient characteristics of speed, torque and line current, when the motor is accelerated with the rated load, are presented in Figures 14–16 for models BM and M2, respectively. The motor is supplied with network voltage and accelerated with rated load of 14 Nm coupled to the motor shaft and load inertia of 0.0066 kgm^2.

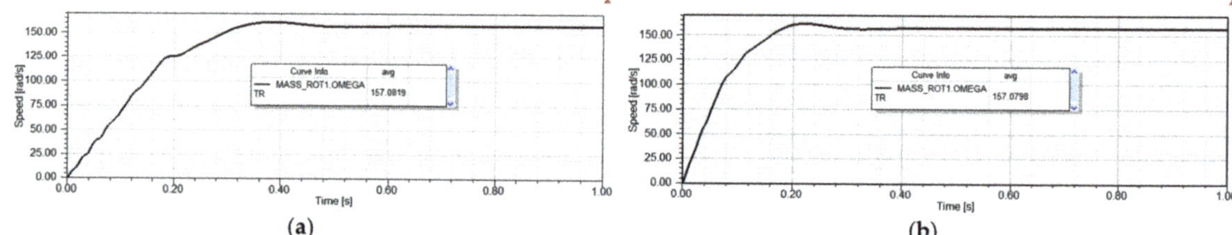

Figure 14. Transient characteristics of speed-load torque 14 Nm and inertia 0.0066 kgm^2 (**a**) BM (**b**) M2.

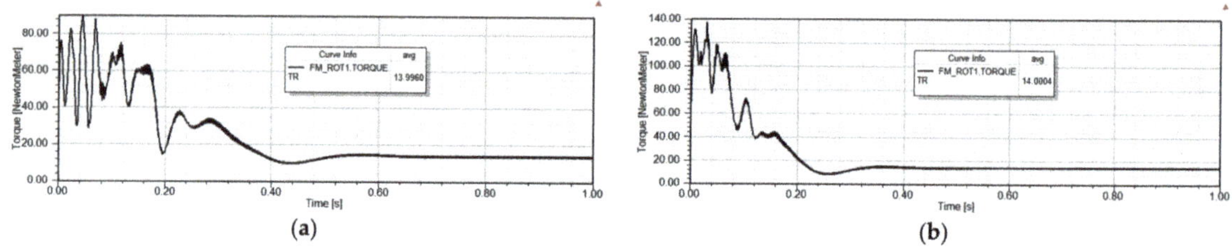

Figure 15. Transient characteristics of torque-load torque 14 Nm and inertia 0.66 kgm^2 (**a**) BM (**b**) M2.

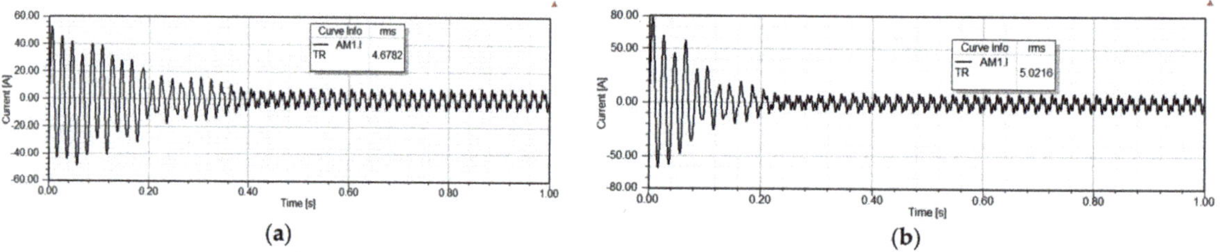

Figure 16. Transient characteristics of current-load torque 14 Nm and inertia 0.66 kgm^2 (**a**) BM (**b**) M2.

The M2 model is simulated for various loads and load inertia. In Figure 17 is presented the characteristic of speed of acceleration of M2 with load torque of 14 Nm and load inertia of 0.37 kgm^2, and with 10 Nm and load inertia of 0.24 km^2. For the above-mentioned loads and moments of inertia, the characteristics of torque and current of M2 are presented in Figures 18 and 19. The maximum load inertia allowed for successful starting is 0.37 kgm^2 for M2 and 0.17 kgm^2 for BM.

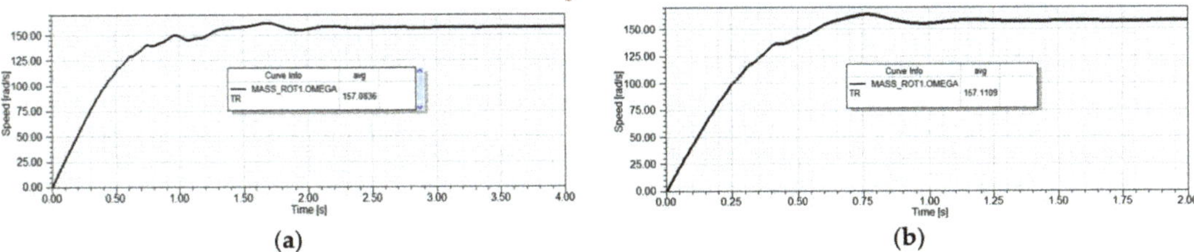

Figure 17. Characteristics of speed of M2 (**a**) load 14 Nm, inertia 0.37 kgm^2 (**b**) load 10 Nm, inertia 0.24 kgm^2.

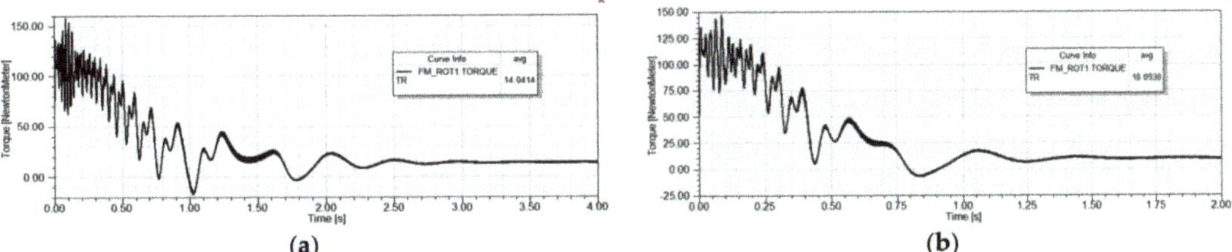

Figure 18. Characteristics of torque of M2 (**a**) torque 14 Nm, inertia 0.37 kgm^2 (**b**) torque 10 Nm, inertia 0.24 kgm^2.

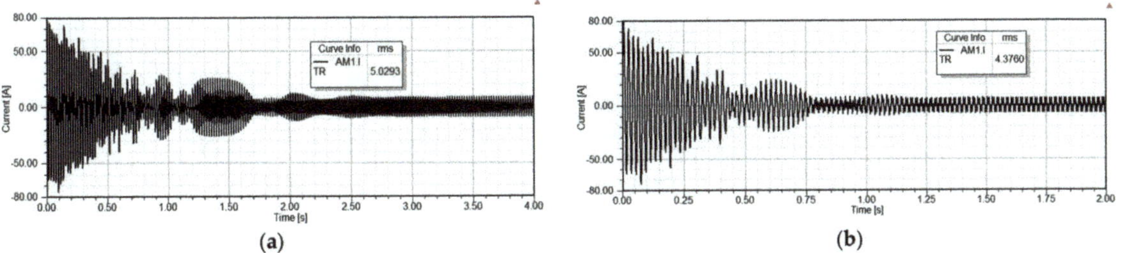

Figure 19. Transient characteristics of current of M2 (**a**) load torque 14 Nm, inertia 0.37 kgm^2 (**b**) load torque 10 Nm, inertia 0.24 kgm^2.

4. Discussion

Obtaining the optimal motor design is not always a straightforward solution, considering that there are many design parameters that have an impact on motor operating characteristics. Improving one operating characteristic may result in deterioration of another. Therefore, four motor parameters (CPS, ORD, MW and MT) that have an impact on motor transient and steady-state characteristics are varied within the prescribed limits, which are determined by designers' experience in order to find the best combination of these four variables to allow obtaining a high efficiency, power factor, and overloading capability along with a cost effective solution regarding material consumption. The line-start synchronous motor with interior asymmetric permanent magnet array is derived from a three-phase squirrel cage motor based on data and the steel laminations from the producer Rade Končar, i.e., the BM model. The BM model is derived from the model of the asynchronous motor (AM) by adding the permanent magnets and flux barriers inside the rotor. Since the overloading capability of the synchronous motor should have a satisfactory value, the air gap length is increased when modifying the AM into BM. In the BM model, the number of conductors per slot was decreased compared to the AM model, which resulted in a lower stator current and considerably lower copper losses in the stator winding. The decrease in the current at BM is also a result of significantly improved power factor at synchronous motor (BM) compared to the asynchronous motor AM. The rotor copper losses at the rated load operation are not present in the BM model of line-start synchronous motor due to its principle of operation. No current is induced at synchronous speed of the motor in the rotor winding; therefore, no copper losses are present in the rotor winding of the synchronous motor. The detailed breakdown of all losses of the both models of the motor, the asynchronous (AM) and the synchronous (BM) are presented in Tables 1 and 4. The above-mentioned modifications of the BM model compared to AM model resulted in a significant increase in the efficiency from 78.4% at AM to 94.3 at BM. The first modification of the motor design which involved only a redesign of the rotor slots were in model M1. The reason for redesigning the rotor slots was to provide more space for magnets in the rotor since the original BM model had a low overloading capability of 1.6 and a maximum output power of 3572 W. By modifying the rotor slots, the magnet thickness and width can be increased, which in turn provides the larger overloading capability of the model of 4326 W or 1.9. The increase in the overloading capability is due to the increased weight of magnet material; consequently, the costs of production are increased as well. No significant improvement of efficiency factor can be observed in the M1 model, compared to the BM model, although the power factor is improved, the line current is decreased and so are the copper losses (Table 4). The magnet thickness and width along with outer rotor diameter, i.e., the air gap length and the number of conductors per slot, are selected as parameters to be varied in the optometric analysis, which resulted in the models M2, M3 and M4. In terms of the overloading capability, efficiency and power factor, the M4 model has the best operating characteristics of efficiency 96.1%, power factor of 0.93 and overloading capability of 2.8, i.e., a maximum output power of 6158 W. The consumption of permanent magnet material is considerable; therefore, the M4 is not the most cost effective solution in terms of material consumption. The models M2 and M3 have similar operating characteristics (efficiency 96 and 95.9, power factor 0.9 and 0.94, and maximum output power of 4929 and 4837, respectively). In terms of permanent magnet consumption, M2 has smaller consumption, 0.69 kg versus 0.8 kg at M3. The M2 model is chosen for further analysis as it has the better efficiency, overloading capability and smaller permanent magnet consumption than the M3 model. M2 also has the smaller air gap, fewer conductors per slot than M1 and, consequently, lower copper losses and greater efficiency than M1. Lower copper losses in the M2 are also a result of the decreased air gap in the M2 model compared to the M1 model, which resulted in the improved power factor; consequently, the current is decreased and, finally, the copper losses are decreased. The fewer conductors per slot, combined with the smaller air gap length, contributed to the lower stator winding resistance, greater power factor, lower current, smaller copper losses and greater efficiency of the M2 model compared to the M1 model. The M2 model also has the modified rotor slot compared to the M1 model. This modification provides more space for magnets in the rotor, which contributes to the greater power factor, lower motor current, smaller copper losses and better efficiency of the M2 model. The modification of the rotor slot is also important for the overloading capability of the motor. The increase in the amount of the magnet material (more available space for magnets in the rotor) increases the overloading capability of this type of line-start synchronous motor. The motor optimization and modification should be evaluated in terms of complete spectrum of operating characteristics, not just in terms of the efficiency or the power factor. Therefore, the rotor slot modification is important for the overloading capability of the motor and this can be clearly observed from the presented data of the M4 model in Table 4. All motor models are calculated with the same steel laminations. The type of steel in the lamination does not have such a drastic impact on the efficiency. The authors' preliminary analysis showed that various types of steel affect the efficiency by no more than two to three percent (depending of the type of the steel and its specific losses). Further research can be extended with the detailed analysis of the impact of the type of the steel laminations and their specific losses on the motor efficiency. The impact of CPS on motor efficiency is presented in Figure 9. The program adjusts the wire diameter according to the CPS in order not to exceed the limit of slot fill factor of 75%. The wire diameter has an impact on the stator winding resistance and, consequently, on the line current, the copper losses and the efficiency. The impact of CPS on the stator winding resistance, line current, copper losses and total losses is presented in Figure 20.

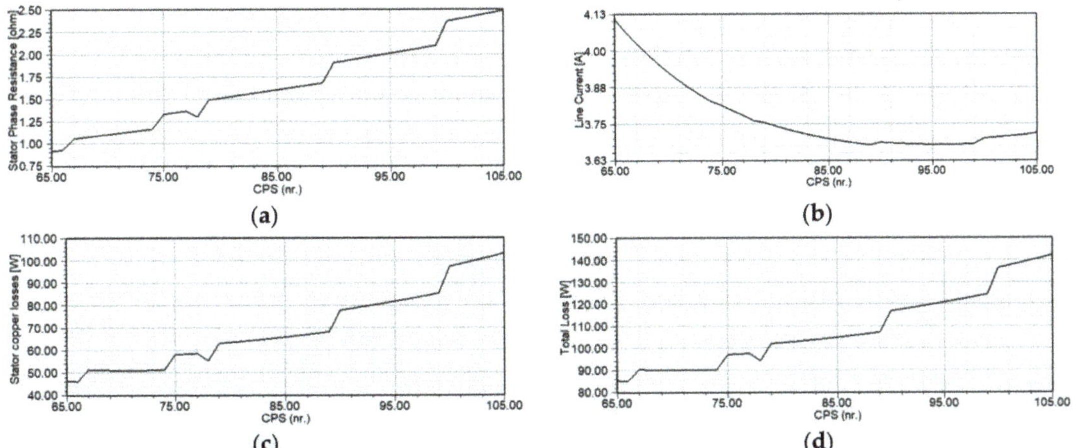

Figure 20. Impact of CPS on (**a**) stator winding resistance (**b**) line current (**c**) stator winding copper losses (**d**) total losses.

The impact of CPS on the stator winding phase resistance is significant and it contributes greatly to the copper losses and, consequently, to the total losses and the efficiency factor (Figure 20c). The impact of CPS on line current is not as pronounced as it is on the winding resistance (Figure 20b). The cooper losses which are significant part of the motor total losses are determined by the impact of CPS on stator winding resistance. The decrease of the number of CPS has a positive impact on the decrease of winding resistance and has a negative impact on the stator current since the current increases and consequently the copper losses as well. Yet, the decrease in the winding resistance and its contribution to the copper losses and consequently to the total losses is more pronounced than the impact of the increase of the line current on the copper losses and consequently to the total losses. Therefore, the decrease in the number of CPS has a positive impact on the improvement of total losses and results in an increase in the efficiency factor. This statement is verified by the data presented in Table 4 (model M2). Figure 20 should support presented result of impact of CPS on the efficiency (Figure 9a). The smaller number of CPS has a positive impact on the increase of efficiency but simultaneously decreases the power factor. Additionally, the smaller number of CPS contributes to the larger starting torque and the overloading capability of the motor. The CPS has a significant impact on all evaluated motor operating characteristics (efficiency, power factor, stating torque and maximum output power). The impact of ORD is more pronounced on the efficiency and the power factor, while it is not the case with the starting torque and the overloading capability (Figure 10). The larger air gap or the smaller ORD decreases the efficiency and the power factor. MW has no significant impact on the efficiency, the power factor, the starting torque and the overloading capability (Figure 11). On the other hand, MT has the greatest impact on motor overloading capability and power factor. The thicker the magnets are, the greater the power factor and the overloading capability are (Figure 12). From all four varied parameters, CPS has the biggest impact on all four analyzed operating characteristics, efficiency, power factor, starting torque and overloading capability. ORD, i.e., air gap length, has a significant impact only on efficiency and the power factor. MW is not a critical parameter regarding analyzed operating characteristics but MT has crucial impact on the overloading capability and the power factor. The impact of varied parameter CPS, ORD, MT and MW on motor efficiency can be observed from Figures 9a, 10a, 11a and 12a). From the data presented in the above-mentioned figures, it can be concluded that variation of CPS had the biggest impact on the percentage of motor efficiency, i.e., the variation of CPS from 65 to 108 impacts the change of efficiency from 93.5 to 96.2%. The change of air gap length has also a significant impact on efficiency, which varies from 94.9% to 96.1% with the increase in the ORD, or with the decrease in the air gap length from 1 mm to 0.5 mm. The impact of magnet width on motor efficiency is not so pronounced. The efficiency percentage varies very little with the variation of magnet width, i.e., from 96.04 % to 96.07 %. The impact of magnet thickness on efficiency is more significant, i.e., it increases from 96.04% to 96.14% with the increase in the magnet thickness. From the above, it is evident that various parameters have different impacts on various motor operating characteristics; therefore, it is necessary to employ computational techniques, such is the case with optometric analysis, which involve fast and accurate calculation of various motor models that allow for the determination of optimum values of analyzed motor parameters which produce the best operating characteristics in terms of complete specter of them, combined with the most cost effective solutions regarding material consumption.

The M2 model has been chosen as an optimal solution regarding operating characteristics and the permanent magnet material consumption. Both models BM and M2 are modeled with FEM for the flux density distributions. According to Figure 13, models BM and M2 have higher flux density in the stator yoke. This can be improved by redesigning the stator, i.e., increasing the stator outer diameter. In this paper, modifications of the motor are performed on the basis of a three-phase squirrel cage motor, a product of Končar, without any changes in the motor outer dimensions.

The line-start synchronous motor should successfully start and accelerate up to the synchronous speed when it is plugged in the three-phase voltage. The successful starting, acceleration and synchronization are determined by the rotor cage winding and the permanent magnets. Their proper design is vital for the successful operation of the motor. The transient characteristics of speed, current and torque, presented in Figures 14–16, allow analysis of the acceleration and synchronous operation of both models BM and M2. Both models are accelerated with the rated load of 14 Nm coupled to the motor shaft. The M2 model has higher starting torque and lower acceleration time than the BM model. Both motors reach the synchronous speed and maintain the synchronous operation. After the acceleration has finished, the motor torque reaches the value of 14 Nm for both models. This can be expected as they are loaded with the rated torque of 14 Nm. Similar observations can be made for the motor current which reaches the rated load current after the motor has accelerated. The analysis is extended by loading the M2 model with different loads and load inertia, i.e., 10 Nm and load inertia of 0.24 kgm^2 and 14 Nm and load inertia of 0.37 kgm^2. In both cases, the motor synchronizes and maintains the synchronism. Table 7 presents the comparison between results obtained in Table 4 from computer models for calculation of motor parameters and characteristics and the results of speed, torque and current from the transient characteristics (Figures 14–16). The average values of speed and torque are calculated by the dynamical model for the last time interval of characteristics and presented in Figures 14 and 15. The rms value of current for the last time interval of current characteristic is presented in Figure 16.

Table 7. Comparison of results from analytical and dynamic models at rated load.

BM	Analytical Calculation	Dynamic Model
Speed (rpm)	1500	1500
Torque (Nm)	14	13.9
Current (A)	3.9	4.7
M2		
Speed (rpm)	1500	1500
Torque (Nm)	14	14
Current (A)	3.9	5

From the results presented in Figure 16, it can be concluded that there is a significant distortion of current waveform due to the presence of harmonics. Harmonics are often present in the current of line-start synchronous motor as a result of permanent magnets inside the rotor. The current higher harmonics cause absorption of distortion power and increase in stray load losses [32]. The rms value of current of the dynamic model has higher value than the analytically calculated current (Table 3) due to presence of harmonics. These harmonics cause the heating effects in the conductors, as the eddy losses are proportional to the square of the frequency. Moreover, harmonics can cause interference in the protection systems, communication systems, and signaling circuits due to electromagnetic induction [33]. The harmonics contribute also to the increased noise and vibrations during motor operation. To mitigate the problem with harmonics some authors propose the usage of filters or modification of rotor teeth width [33,34].

From the results presented in Table 7, it can be concluded that computer model for analytical calculation of parameters and characteristics and the dynamic model have satisfactory similarity of results of speed, current and torque. In addition to the verification of the motor dynamic regimes, the presented results in Table 6 should verify the accuracy of both models, the computer model for calculating parameters, the steady-state characteristics and the dynamic model.

The M2 model achieves high efficiency and a very good power factor combined with good overloading capabilities, considerably higher than the efficiency of motor for the same power rating found in [4]. Yet, the derived model is theoretical, based on computer simulations. The proposed model should be verified by the prototype and experimental measurements. It can be expected that the manufactured model will have lower efficiency and power factor. The model is subject to manufacturing limitations such as achieving a good slot fill factor, manufacturing tolerances regarding length of the air gap, built-in material in the motor construction and accurate measurement of frictional and windage losses.

5. Conclusions

The line-start synchronous motor has drawn the attention of scientists and industry professionals as a possible replacement of three-phase squirrel cage induction motors, especially in constant speed applications, due to their high efficiency and good power factor. The strict regulations of the EU market regarding usage of IE3 efficiency class of motors (that can be achieved with three-phase squirrel cage motors with numerous modifications that require more material with high quality and low losses) increases the interest for these line-start synchronous motors, which can easily achieve IE4 efficiency class. Among the line-start synchronous motors, there are various topologies which require different designs of the rotors and produce different motor operating characteristics. The authors have chosen to analyze the line-start synchronous motor with an interior permanent magnet asymmetric array topology as this configuration needs more detailed analysis and modification of rotor design, due to the availability of the space in the rotor to place magnets with sufficient dimensions, in order to achieve good efficiency, power factor and overloading capability of the motor. Starting from the three-phase squirrel cage motor of 2.2 kW, the first model (BM) is obtained by placing the magnets inside the rotor without any other modifications in the motor. The obtained BM model has high efficiency and power factor but relatively low overloading capability. The second model (M1) with modified rotor slots and magnet dimensions has high efficiency and power factor, improved overloading capability of the motor but relatively high consumption of permanent magnet material. Therefore, more modifications of motor design were needed that included modification of the air gap length, magnet width and thickness and the number of the conductors per slot. In order to determine the best combination of these four parameters that produce the high power factor and efficiency, with good overloading capability and low consumption of permanent magnet material, the optometric analysis was run and more than 25,257 combinations were solved. This analysis is a useful tool as the increase in one parameter can improve one operating characteristic and worsen the other, and vice versa. When four different parameters are varied within certain boundaries simultaneously, without optometric analysis, it is very difficult to determine which value each of these parameters should have that will produce a model of the motor which will satisfy four various operating characteristics regarding optimal or improved operation in comparison to the starting model. Among these numerous combinations, the M2 model, with a sufficiently high power factor, efficiency and overloading capability (higher efficiency and overloading factor than models BM and M1) and considerably lower consumption of permanent magnet material is chosen as the optimal solution. The impact of each varied parameter on motor operating characteristics is analyzed, providing detailed insight into which design guidelines should be followed for obtaining satisfactory design of the motor. The chosen model (M2) was analyzed with FEM for the magnetic flux density distribution and with dynamic models for obtaining the transient characteristics. The M2 model has some areas of stator yoke with high flux density which can be improved by increasing the stator diameter, subject to further research. The dynamic behavior of M2 is satisfactory since the motor reaches the synchronous speed and continues with stable operation.

The careful analysis of each parameter and its impact on motor operating characteristics can significantly improve the motor operation, leading to the cost effective design of the motor. The proposed model is based on computer analysis and simulations. Its prototyping is highly affected by various manufacturing details such as obtaining a good slot fill factor or vibrations and noise as a result of the air gap length, which can have a significant impact on the final outcomes of this analysis.

Author Contributions: Conceptualization, V.S.; methodology, V.S., P.J. and D.M.; software, V.S.; validation, V.S., D.M. and P.J.; formal analysis, V.S.; investigation, V.S.; resources, V.S.; data curation, V.S.; writing—original draft preparation, V.S.; writing—review and editing, P.J.; visualization, V.S.; supervision, P.J.; project administration, D.M.; funding acquisition, D.M. and P.J. All authors have read and agreed to the published version of the manuscript.

Funding: This publication was created thanks to support under the Operational Program Integrated Infrastructure for the project: International Center of Excellence for Research on Intelligent and Secure Information and Communication Technologies and Systems-II. stage, ITMS code: 313021W404, co-financed by the European Regional Development Fund.

Conflicts of Interest: The authors declare no conflict of interest. The funders had no role in the design of the study; in the collection, analyses, or interpretation of data; in the writing of the manuscript, or in the decision to publish the results.

Nomenclature

List of used symbols:

i_{dr}	Rotor d-current
i_{qr}	Rotor q-current
i_{ds}	Stator d-current
i_{qs}	Stator q-current

L	Inductance
R_s	Stator resistance
T	Motor torque
u	Voltage
ω_s	Synchronous speed
ω_m	Motor speed
Ψ	Flux linkage
I_{dgmrm}	Maximum permitted value of steady-state current before demagnetization (A)
p	Number of poles
μ_0	magnetic permeability of vacuum
K_{w1}	Winding factor
N_c	number of turns per phase of stator winding
Br	Residual flux density in T
B_d	Corresponding flux density at H_d at which point the magnetic polarization vector M collapses

List of subscripts:

dr	Rotor d-axis
ds	Stator d-axis
m	Magnet
md	Mutual d-axis
mq	Mutual q-axis
qr	Rotor q-direction
qs	Stator q-direction
r	Rotor
s	Stator

References

1. Siemens. Available online: https://press.siemens.com/global/en/pressrelease/high-efficiency-motor-series-siemens-now-consistently-available-efficiency-class-ie4 (accessed on 18 October 2021).
2. WEG. Available online: https://www.quantum-controls.co.uk/wp-content/uploads/2020/11/w22-three-phase-motor-1.pdf (accessed on 18 October 2021).
3. Verucchi, C.; Ruschetti, C.; Giraldo, E.; Bossio, G.; Bossio, J. Efficiency optimization in small induction motors using magnetic slots. *Electron. Power Syst. Res.* **2017**, *152*, 1–8. [CrossRef]
4. WEG. Available online: https://static.weg.net/medias/downloadcenter/h01/hfc/WEG-w22-quattro-european-market-50025713-brochure-english-web.pdf (accessed on 18 October 2021).
5. Alibad, A.D.; Mirsalim, M.; Ershad, N.F. Line-Start Permanent-Magnet Motors: Significant Improvements in Starting Torque, Synchronization, and Steady-State Performance. *IEEE Trans. Magn.* **2010**, *46*, 4060–4072. [CrossRef]
6. Knypiński, Ł. Optimal design of the rotor geometry of line-start permanent magnet synchronous motor using bat algorithm. *Open Phys.* **2017**, *15*, 965–970. [CrossRef]
7. Gwozdziewicz, M.; Zalas, P.; Zawilak, J. Starting process of medium power line start permanent magnet synchronous motor. *Prezglad Elektrotechniczney* **2017**, *93*, 62–64. [CrossRef]
8. Wymeersch, B.; De Belie, F.; Rasmussen, C.; Vandevelde, L. Classification Method to Define synchronization Capability Limits of Line-Start Permanent-Magnet Motor Using Mesh-Based Magnetic equivalent Circuit Computation Results. *Energies* **2018**, *11*, 998. [CrossRef]
9. Chama, A.; Sorgdrager, J.A.; Wang, R.-J. Analytical Synchronization Analysis of Line-Start Permanent Magnet Synchronous Motor. *Progress Electromagn. Res. M* **2016**, *48*, 183–193. [CrossRef]
10. Qiu, H.; Zhang, Y.; Hu, K.; Yang, C.; Yi, R. The Influence of Stator Winding Turns on the Steady-State Performance of Line-Start Permanent Magnet Synchronous Motors. *Energies* **2019**, *12*, 2363. [CrossRef]
11. Baranski, M.; Szelag, W.; Lyskawinski, W. An analysis of a start-up process in LSPMSMs with aluminum and copper rotor bars considering the coupling of electromagnetic and thermal phenomena. *Arch. Electr. Eng.* **2019**, *68*, 933–946.
12. Baranski, M.; Szelag, W.; Lyskawinski, W. Analysis of the Partial Demagnetization Process of Magnets in a Line Start Permanent Magnet Synchronous Motor. *Energies* **2020**, *13*, 5562. [CrossRef]
13. Zöhra, B.; Akar, M.; Eker, M. Design of a Novel Line Start Synchronous Motor Rotor. *Electronics* **2019**, *8*, 25. [CrossRef]
14. Jędryczka, C.; Knypiński, Ł.; Demenk, L.; Sykulski, K.J. Methodology for Cage Shape Optimization of a Permanent Magnet Synchronous Motor Under Line Start Conditions. *IEEE Trans. Magn.* **2018**, *54*, 8102304. [CrossRef]
15. Knypiński, Ł.; Pawełoszek, L.; Le Menach, Y. Optimization of Low-Power Line-Start PM Motor Using Gray Wolf Metaheuristic Algorithm. *Energies* **2020**, *13*, 1186. [CrossRef]

16. Knypiński, Ł.; Nowak, L. The algorithm of multi-objective optimization of PM synchronous motors. *Przeglad Elektrotechniczney* **2019**, *95*, 242–245. [CrossRef]
17. Sorgdrager, A.J.; Wang, R.-J.; Pfeffer, A.K. Optimization of a line-start permanent magnet synchronous machine for a load specific application. In Proceedings of the 23rd Southern African Universities Power Engineering Conference, Johannesburg, South Africa, 28–30 January 2015; pp. 216–220.
18. Sorgdrager, A.J.; Smith, R.; Wang, R.-J. Rotor design of a line start permanent magnet synchronous machine using Taguchi method. In Proceedings of the 23rd Southern African Universities Power Engineering Conference, Johannesburg, South Africa, 28–30 January 2015; pp. 227–232.
19. Sarac, J.V.; Stefanov, G. Various Rotor Topologies of Line-Start Synchronous Motor for Efficiency Improvement. *Power Electron. Drives* **2020**, *5*, 83–95. [CrossRef]
20. Koncar-Mes. Available online: https://koncar-mes.hr/wp-content/uploads/2020/08/Electric-motors.pdf (accessed on 16 October 2021).
21. Duan, Y. Method for Design and Optimization of Surface Mount Permanent Magnet Machines and Induction Machines. Ph.D. Dissertation, Georgia Institute of Technology, Atlanta, GA, USA, 2010. Available online: https://smartech.gatech.edu/bitstream/handle/1853/37280/duan_yao_201012_phd.pdf (accessed on 18 October 2021).
22. Karami, M.; Mariun, N.; Mehrjou, R.M.; Ab Kadir, A.Z.M.; Misron, N.; Radzi, M.A.M. Static Eccentricity Fault Recognition in Three-Phase Line Start Permanent Magnet synchronous Motor Using Finite Element Method. *Math. Probl. Eng.* **2014**, *2014*, 132647. [CrossRef]
23. Vukotić, M.; Rodić, D.; Benedičič, B.; Miljavac, D. Cogging torque in slotless permanent magnet machines. *J. Electr. Eng.* **2020**, *71*, 195–202. [CrossRef]
24. Behbahanifard, H.; Sadoughi, A. Cogging Torque Reduction in Line-Start Permanent Magnet Synchronous Motor. *J. Electr. Eng. Technol.* **2016**, *11*, 878–888. [CrossRef]
25. Ogbuka, C.; Nwosu, C.; Agu, M. Dynamic and steady state performance comparison of line-start permanent magnet synchronous motors with interior and surface rotor magnets. *Arch. Electr. Eng.* **2016**, *65*, 105–116. [CrossRef]
26. Sriprang, S.; Nahid-Mobarakeh, B.; Takorabet, N.; Pierfederici, S.; Kumam, P.; Bizon, N.; Taghavi, N.; Vahedi, A.; Mungporn, P.; Thounthong, P. Design and control of permanent magnet assisted synchronous reluctance motor with copper loss minimization using MTPA. *J. Electr. Eng.* **2020**, *71*, 11–19. [CrossRef]
27. Ansys Inc. *Maxwell 2D User's Guide 2010*; Ansys Inc.: Canonsburg, PA, USA, 2020; p. 11-3-36.
28. Liber, F.; Soulard, J.; Engström, J. Design of a 4-Pole Line Start Permanent Magnet Synchronous Motor. Available online: https://citeseerx.ist.psu.edu/viewdoc/download?doi=10.1.1.529.5949&rep=rep1&type=pdf (accessed on 17 October 2021).
29. Krause, W.O.; Sudhoff, S.D. *Analysis of Electrical Machinery*, 2nd ed.; IEEE Press: New York, NY, USA, 1995; p. 170.
30. Zubayer, M.H. Design Analysis of Line-Start Interior Permanent Magnet Synchronous Motor. 2011. Available online: https://research.library.mun.ca/10043/1/Zubayer_HossanM.pdf (accessed on 2 November 2021).
31. Elistratova, V. Optimal Design of Line-Start Permanent Magnet Synchronous Motors of High Efficiency. Electric Power. Ecole Centrale de Lille. 2015. Available online: https://tel.archives-ouvertes.fr/tel-01308575/document (accessed on 3 November 2021).
32. Zawilak, T.; Zawilak, J. Minimization of Higher Harmonics in Line-Start Permanent Magnet Synchronous Motor. *Prace Naukowe Instytutu Maszyn, Napędów i Pomiarów Elektrycznych Politechniki Wrocławskiej Studia i Materiały* **2009**, *63*, 107–115. Available online: http://yadda.icm.edu.pl/yadda/element/bwmeta1.element.baztech-article-BPW9-0009-0014 (accessed on 20 October 2021).
33. Gwoździewicz, M.; Antal, L. Investigation of Line Start Permanent Magnet Synchronous Motor and Induction Motor Properties. *Prace Naukowe Instytutu Maszyn Napędów i Pomiarów Elektrycznych Politechniki Wrocławskiej Studia i Materiały* **2010**, *64*, 13–20. Available online: http://yadda.icm.edu.pl/yadda/element/bwmeta1.element.baztech-article-BPW6-0021-0014 (accessed on 20 October 2021).
34. Chingale, G.S.; Ugale, R.T. Harmonic Filter Design for Line Start Permanent Magnet Synchronous Motor. In Proceedings of the International Conference on Advances in Electrical Engineering (ICAEE), Vellore, India, 9–11 January 2014. [CrossRef]

MDPI

Article

Optimization of Fuzzy Controller for Predictive Current Control of Induction Machine

Toni Varga *,†, Tin Benšić †, Marinko Barukčić and Vedrana Jerković Štil

Faculty of Electrical Engineering, Computer Science and Information Technology Osijek, Kneza Trpimira 2b, 31000 Osijek, Croatia; tin.bensic@ferit.hr (T.B.); marinko.barukcic@ferit.hr (M.B.); vedrana.jerkovic@ferit.hr (V.J.Š.)

* Correspondence: toni.varga@ferit.hr

† These authors contributed equally to this work.

Abstract: An optimization procedure for type 1 Takagi–Sugeno Fuzzy Logic Controller (FLC) parameter tuning is shown in this paper. Ant colony optimization is used to obtain the optimal controller parameters, and only a small amount of post-optimization parameter adjustment is needed. The choice of controller parameters is explained, along with the methodology behind the criterion for objective function value calculation. The optimized controller is implemented as an outer-loop speed controller for Predictive Current Control (PCC) of an induction machine. The performance of the proposed control method is compared with that of several other model predictive control methods. The results show a 55% decrease in speed tracking error and 74% decrease in torque overshoot.

Keywords: induction machine; ant colony optimization; predictive current control; fuzzy logic control; Takagi–Sugeno

Citation: Varga, T.; Benšić, T.; Barukčić, M.; Jerković Štil, V. Optimization of Fuzzy Controller for Predictive Current Control of Induction Machine. *Electronics* **2022**, *11*, 1553. https://doi.org/10.3390/electronics11101553

Academic Editor: Bor-Ren Lin

Received: 21 April 2022
Accepted: 10 May 2022
Published: 12 May 2022

Publisher's Note: MDPI stays neutral with regard to jurisdictional claims in published maps and institutional affiliations.

1. Introduction

Induction machines play a crucial role in modern industry, ranging from simple applications such as driving fans or pumps to more precise usages such as conveyor belts or plastic injection molding [1]. Conventional induction machine control methods such as Field-Oriented Control (FOC) [2] and Direct Torque Control (DTC) [3] can suffer from sensitivity to parameter and torque changes [4], or torque and flux pulsations and control problems at low speeds [5]. For high-performance drives, which require a fast dynamic response and disturbance rejection, these shortcomings must be improved, or new methods must be investigated. One direction of research is incorporating fuzzy logic into conventional control structures to improve the dynamics.

Research in [6] shows an increasing trend of computational intelligence implementation in control applications. An overview of the recent literature proves the research interest in fuzzy logic in drive systems. The authors of [7] investigate the implementation of FLC with a reduced computation burden. In [8], the authors incorporate FLC into a classic DTC structure to improve the torque ripple of the high-performance drive. The authors of [9] investigate FLC for low-speed induction machine operation. In [10], the authors investigate the computation burden of FLC depending on the size of the fuzzy rules table. In [11,12], type 2 FLC is investigated to improve the DTC structure on five- and three-level inverters. In [13,14], the authors use FLC to improve the classic DTC in an induction machine drive with a two-level converter and dual-stator machine drive, respectively. The authors of [15] improve the FOC structure for an induction machine by developing a fuzzy speed controller with an algorithm for automatic gain output adjustment. Tir et al. use fuzzy logic to improve the FOC structure for a single system in [16] and a multi-machine drive system in [17]. Bessaad et al. use a fuzzy system in [18] for developing a correction regulator in a multi-machine system with a single inverter supply. The authors of [19] develop a custom search algorithm for induction machine fuzzy-PI controller tuning. To improve efficiency, the authors of [20,21] use fuzzy logic for the online search of the minimum power losses of an induction machine drive and self-excited induction generator, respectively. In [22], the authors use fuzzy logic to develop an expert system for induction machine fault diagnosis. The authors of [23] use FLC to calculate a suitable voltage vector for a five-phase induction machine to reduce torque ripple. The authors of [24] use fuzzy logic as a decision mechanism for weighing factor selection in predictive torque control. In [25], the authors incorporate two FLCs as inputs to a feedback linearization algorithm to improve drive dynamics. Saghafinia et al. use FLC in [26] as a speed controller in a sliding mode control structure. In [27], the author presents a fuzzy-PI controller utilized for minimizing energy consumption while maximizing the performance of an induction machine drive. In [28], Youb et al. develop a fuzzy-PI controller for a dual-star induction machine with online adaptation of proportional and integral gains.

There are a number of research papers showing the usage of fuzzy systems in parameter estimation. The authors of [29] use fuzzy logic concepts for estimation purposes by building an observer using the Takagi–Sugeno model of an induction machine. Jabbour et al. use fuzzy logic in [30] for online parameter estimation. In [31], the authors utilize a type 1 and type 2 fuzzy controller for a model reference adaptive system and compare the results. The authors of [32] build a Luenberger observer using a fuzzy logic system that outputs state variable estimates. In an older paper [33], the authors optimize a fuzzy system for the purpose of estimating the rotor time constant. In [34], Shukla et al. use fuzzy logic for the model reference speed adaptation of an induction machine that is controlled via DTC.

Fuzzy logic is also used in power and frequency control, as can be seen in papers such as [35], where the authors use fuzzy logic to tune the proportional, integral, and derivative gains of a doubly fed induction generator (*DFIG*) PID controller. In [36], the authors also use fuzzy concepts to improve the DTC of a DFIG, while in [37], FLC is used for active power control. In [38], Dewangan et al. replace a classic PI controller with FLC to improve the performance of a wind-driven self-excited induction generator during fault and variable wind speed conditions. The authors of [39] develop FLC for a six-phase induction generator that shows superiority over the classic controller in fault conditions.

In this paper, the authors present an optimization procedure for the fuzzy speed controller that enhances the speed tracking and torque response of the induction machine, which is controlled using the PCC algorithm. In recent years, a few papers have dealt with this problem in the induction machine drive field. In [40], George et al. use an optimization procedure to optimize the speed control of an electric vehicle. Similarly, the authors of [41] use an optimization approach to optimize frequency control in multi-area interconnected power systems. Both papers offer several criteria for objective function value calculation. In [37], the authors use the integral time squared error to calculate the objective function value for the particle swarm optimization algorithm, which is used to optimize DFIG power control. The authors of [42] optimize FLC for an induction machine and use the mean average error as a criterion to optimize seven membership functions for inputs and output while using the Mamdani-type defuzzification process. In this paper, the authors use five membership functions for inputs, and a Takagi–Sugeno-type defuzzification process. Several criteria for objective function value calculation are investigated and results show improvements upon the classic speed controller. The paper is organized as follows: In Section 2, the dynamic model of the induction machine is presented and a control overview is given. In Section 3, the fuzzy logic speed controller that is being optimized in this paper is presented and each part of the controller is described. The problem statement and optimization procedure (optimizer, controller parameters, and the choice of objective function) are presented in Section 4. In Section 5, results for different optimization approaches are displayed and discussed. Control system performance is shown in Section 6, while the comparison with several model predictive methods is carried out in Section 7. In Section 8, the authors discuss caveats and future research possibilities.

2. Induction Machine Model and Control Overview

The dynamic model of the induction machine is written in the stationary $\alpha\beta$-reference frame and it is shown by Equations (1)–(4), where J is the inertia constant of the machine, ω_r is the rotor shaft speed, p is the number of pole pairs, T_e is electromechanical torque, and T_l is load torque.

$$\boldsymbol{v}_{s\alpha\beta} = \boldsymbol{R}_s \boldsymbol{i}_{s\alpha\beta} + \frac{\mathrm{d}}{\mathrm{d}t}\boldsymbol{\psi}_{s\alpha\beta} \tag{1}$$

$$0 = \boldsymbol{R}_r \boldsymbol{i}_{r\alpha\beta} - \boldsymbol{J}_r p\omega_r \boldsymbol{\psi}_{r\alpha\beta} + \frac{\mathrm{d}}{\mathrm{d}t}\boldsymbol{\psi}_{r\alpha\beta} \tag{2}$$

$$J\frac{\mathrm{d}}{\mathrm{d}t}\omega_r = T_e - T_l \tag{3}$$

$$T_e = \frac{3}{2}p(\psi_{s\alpha} i_{s\beta} - \psi_{s\beta} i_{s\alpha}) \tag{4}$$

Equation (5) displays electrical quantities, where $\boldsymbol{v}_{s\alpha\beta}$ represents the stator voltage vector, $\boldsymbol{i}_{s\alpha\beta}$ and $\boldsymbol{i}_{r\alpha\beta}$ represent the stator and rotor current vector, while $\boldsymbol{\psi}_{s\alpha\beta}$ and $\boldsymbol{\psi}_{r\alpha\beta}$ describe stator and rotor flux linkages.

$$\boldsymbol{v}_{s\alpha\beta} = \begin{bmatrix} v_{s\alpha} \\ v_{s\beta} \end{bmatrix} \quad \boldsymbol{i}_{s\alpha\beta} = \begin{bmatrix} i_{s\alpha} \\ i_{s\beta} \end{bmatrix} \quad \boldsymbol{i}_{r\alpha\beta} = \begin{bmatrix} i_{r\alpha} \\ i_{r\beta} \end{bmatrix} \quad \boldsymbol{\psi}_{s\alpha\beta} = \begin{bmatrix} \psi_{s\alpha} \\ \psi_{s\beta} \end{bmatrix} \quad \boldsymbol{\psi}_{r\alpha\beta} = \begin{bmatrix} \psi_{r\alpha} \\ \psi_{r\beta} \end{bmatrix} \tag{5}$$

Electrical parameters are described by Equation (6), where $\boldsymbol{R}_s$ and $\boldsymbol{R}_r$ are the stator and rotor resistance matrices, while $\boldsymbol{J}_r$ represents the rotation matrix.

$$\boldsymbol{R}_s = \begin{bmatrix} R_s & 0 \\ 0 & R_s \end{bmatrix} \quad \boldsymbol{R}_r = \begin{bmatrix} R_r & 0 \\ 0 & R_r \end{bmatrix} \quad \boldsymbol{J}_r = \begin{bmatrix} 0 & -1 \\ 1 & 0 \end{bmatrix} \tag{6}$$

Additionally, the relationship between fluxes and currents is described by Equations (7) and (8), but for a deeper understanding of the induction machine model, the reader is referred to [43].

$$\boldsymbol{\psi}_{s\alpha\beta} = L_s \boldsymbol{i}_{s\alpha\beta} + L_m \boldsymbol{i}_{r\alpha\beta} \tag{7}$$

$$\boldsymbol{\psi}_{r\alpha\beta} = L_r \boldsymbol{i}_{r\alpha\beta} + L_m \boldsymbol{i}_{s\alpha\beta} \tag{8}$$

Model predictive control has attracted research interest in recent years, as seen in papers such as [44–46]. This being the case, the control method that is being modified in this paper is PCC for an induction machine, which falls into the family of model predictive control structures. In the following text, it is explained how it works.

Firstly, a discrete state space model for computing the current prediction $i_{s\alpha\beta}[k+1]$ is formed by selecting stator currents $i_{s\alpha\beta}[k]$ and rotor fluxes $\psi_{r\alpha\beta}[k]$ as state variables. Equation (9) represents the final expression to calculate the current predictions, where $\sigma = 1 - \frac{L_m^2}{L_s L_r}$ represents the leakage inductance factor, $\tau_r = \frac{L_r}{R_r}$ represents the rotor time constant, and T_s represents the discretization sampling time.

$$\begin{aligned} \begin{bmatrix} i_{s\alpha}[k+1] \\ i_{s\beta}[k+1] \end{bmatrix} = & \begin{bmatrix} 1 - \frac{(R_s + \frac{L_m^2}{L_r \tau_r})T_s}{\sigma L_s} & 0 & \frac{L_m T_s}{\sigma L_s L_r \tau_r} & \frac{L_m p\omega_r T_s}{\sigma L_s L_r} \\ 0 & 1 - \frac{(R_s + \frac{L_m^2}{L_r \tau_r})T_s}{\sigma L_s} & -\frac{L_m p\omega_r T_s}{\sigma L_s L_r} & \frac{L_m T_s}{\sigma L_s L_r \tau_r} \end{bmatrix} \begin{bmatrix} i_{s\alpha}[k] \\ i_{s\beta}[k] \\ \psi_{r\alpha}[k] \\ \psi_{r\beta}[k] \end{bmatrix} \\ & + \begin{bmatrix} \frac{T_s}{\sigma L_s} & 0 \\ 0 & \frac{T_s}{\sigma L_s} \end{bmatrix} \begin{bmatrix} v_{s\alpha}[k] \\ v_{s\beta}[k] \end{bmatrix} \end{aligned} \tag{9}$$

Equation (10) represents the cost function that is used to derive the control law by minimizing the squared error between reference currents $i_\alpha^*[k]$, $i_\beta^*[k]$ and current predictions.

By inserting (9) into (10) and solving (11), the optimal voltage vector $\boldsymbol{v}^*_{\alpha\beta}$ can be obtained to minimize the cost function and drive the machine in the desired state.

$$\boldsymbol{G} = \begin{bmatrix} g_\alpha \\ g_\beta \end{bmatrix} = \begin{bmatrix} (i^*_\alpha[k] - i_\alpha[k+1])^2 \\ (i^*_\beta[k] - i_\beta[k+1])^2 \end{bmatrix} \tag{10}$$

$$\frac{\partial \boldsymbol{G}}{\partial \boldsymbol{v}_{\alpha\beta}} = 0 \tag{11}$$

Equation (12) represents the solution of (11), which is used as a control law.

$$\begin{aligned} \boldsymbol{v}^*_{\alpha\beta}[k+1] = &\begin{bmatrix} R_s + \frac{L_m^2}{L_r\tau_r} - \frac{\sigma L_s}{T_s} & 0 & -\frac{L_m}{L_r\tau_r} & -\frac{L_m p\omega_r}{L_r} \\ 0 & R_s + \frac{L_m^2}{L_r\tau_r} - \frac{\sigma L_s}{T_s} & \frac{L_m p\omega_r}{L_r} & -\frac{L_m}{L_r\tau_r} \end{bmatrix} \begin{bmatrix} i_{s\alpha}[k] \\ i_{s\beta}[k] \\ \psi_{r\alpha}[k] \\ \psi_{r\beta}[k] \end{bmatrix} \\ &+ \begin{bmatrix} \frac{\sigma L_s}{T_s} & 0 \\ 0 & \frac{\sigma L_s}{T_s} \end{bmatrix} \begin{bmatrix} i^*_{s\alpha}[k] \\ i^*_{s\beta}[k] \end{bmatrix} \end{aligned} \tag{12}$$

Reference current value $i^*_\alpha[k]$ is calculated from the desired rotor flux of the machine, while reference current $i^*_\beta[k]$ is generated by the FLC acting as a speed controller. Figure 1 represents the final topology of the control structure, and more details about the PCC structure can be found in [47].

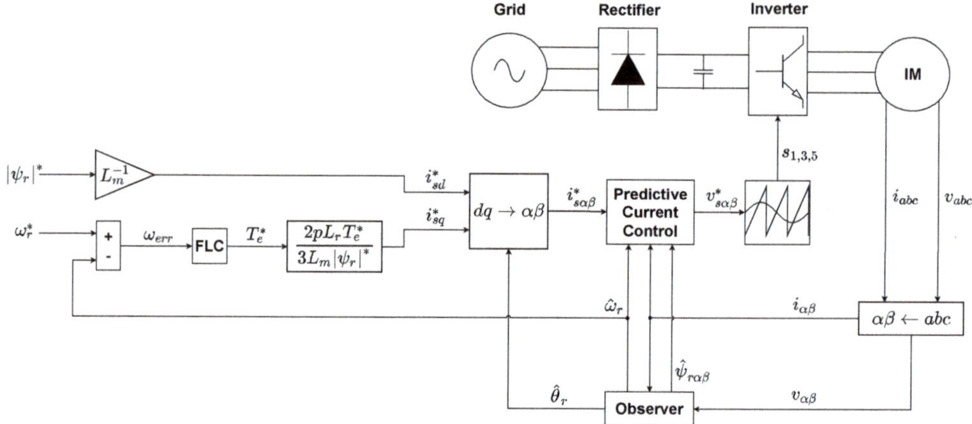

Figure 1. Simulated control method.

3. Fuzzy Logic Controller

The fuzzy inference system block diagram is shown in Figure 2. It represents the computation steps used to transfer inputs to the fuzzy system into control outputs used for generating reference values for induction machine predictive current control. First, input values are turned into fuzzy values by the process of fuzzification. These fuzzy values are calculated using predefined membership functions. The following text explains the selection of input and output variables and membership functions that are to be optimized for induction machine speed control.

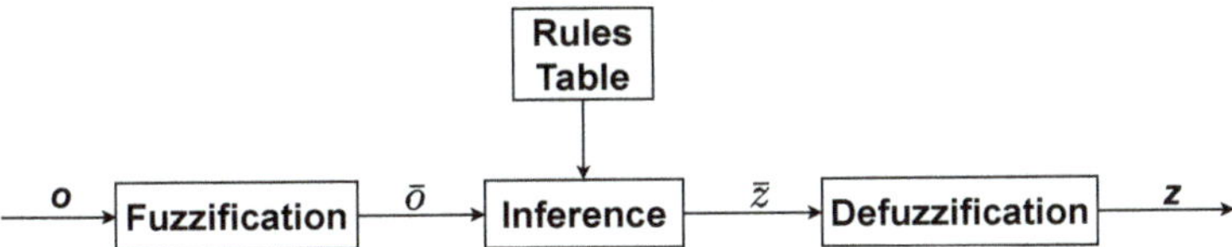

Figure 2. Fuzzy logic controller structure.

3.1. Input and Output of Fuzzy Controller

Input variables to the fuzzy controller are chosen to be the speed tracking error ω_{err} and its derivative $\dot{\omega}_{err}$, while the output is defined as fuzzy error f_{err}. All relevant vectors and sets that will be used are: real valued input vector $\boldsymbol{o}$, input linguistic variable vector $\bar{\boldsymbol{o}}$, and input linguistic value sets $\bar{O}_i$, where i represents the input ordinal number. Regarding the output, z represents the real valued output vector, $\bar{z}$ represents the output linguistic variable vector, and $\bar{Z}$ represents the output linguistic value set. Established vectors and sets are defined as follows:

- $\boldsymbol{o} = \begin{bmatrix}\omega_{err} & \dot{\omega}_{err}\end{bmatrix}$,
- $\bar{\boldsymbol{o}} = [\textit{"Speed Error"} \quad \textit{"Speed Error Derivative"}]$,
- $\bar{O}_1 = \bar{O}_2 = \{\textit{"Negative Big", "Negative Small", "Zero", "Positive Small", "Positive Big"}\}$,
- $z = \begin{bmatrix} f_{err} \end{bmatrix}$,
- $\bar{z} = [\textit{"Fuzzy Error"}]$,
- $\bar{Z} = \{\textit{"Negative", "Zero", "Positive"}\}$.

3.2. Input Membership Functions

Membership functions are structured for ith input linguistic variable $\bar{o}_i$ and kth possible linguistic value $\bar{O}^k$. Their structure depends on the observed linguistic variable and its linguistic value. Equations (13)–(15) represent the membership function definitions for all the possible combinations of index values i and superscript values k. It can be seen that they are unit piecewise linear and defined using only two parameters, a_{ik} and b_{ik}. Figure 3 gives a deeper insight into the membership functions and their parameters.

$$\mu_{\bar{O}_i^k}(o_i) = max\left[\left(1 - \left|\frac{a_{ik} - o_i}{b_{ik}}\right|\right), 0\right], \quad i \in \{1,2\}, \quad k \in \{2,3,4\} \tag{13}$$

$$\mu_{\bar{O}_i^k}(o_i) = \begin{cases} 1, & o_i < a_{ik} \\ 0, & o_i > b_{ik} \\ \frac{a_{ik} - o_i}{b_{ik} - a_{ik}} + 1, & a_{ik} \le o_i \le b_{ik} \end{cases} \qquad i \in \{1,2\}, \quad k = 1 \tag{14}$$

$$\mu_{\bar{O}_i^k}(o_i) = \begin{cases} 0, & o_i < a_{ik} \\ 1, & o_i > b_{ik} \\ \frac{o_i - a_{ik}}{b_{ik} - a_{ik}}, & a_{ik} \le o_i \le b_{ik} \end{cases} \qquad i \in \{1,2\}, \quad k = 5 \tag{15}$$

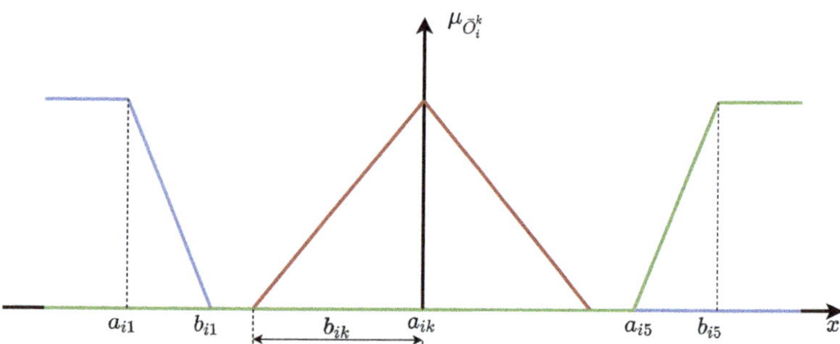

Figure 3. Topology of the fuzzy controller.

The fact that there are only two parameters required for membership function definition offers simplicity of implementation and can be exploited in the optimization procedure of the fuzzy speed controller, as is seen in the following sections. The next step is to obtain the controller output value through the process of fuzzy inference and defuzzification.

3.3. Fuzzy Inference

In control, it is common practice to follow the convention shown in Table 1 of [10] or Table 2 of [40] for fuzzy rule table construction, where an equal number of input and output linguistic values exist in each set $\bar{O}_i$ and $\bar{Z}$. In this paper, however, this convention is not followed since the input linguistic value set contains five elements, while the output linguistic value set consists of three elements. This is done to reduce the number of decision variables in the optimization problem and speed up the overall computation. Table 1 represents the fuzzy rule table that is used in this paper. Abbreviations are used to represent the linguistic values: *N* for *Negative*, *P* for *Positive*, *Z* for *Zero*, *S* for *Small*, and *B* for *Big*. Rules are interpreted as follows: "if Speed Error is Negative-Big and Speed Error Derivative is Negative-Big, then Fuzzy Error is Negative", and so on.

Table 1. Fuzzy Rules.

$\bar{o}_2^m$ \ $\bar{o}_1^n$	NB	NS	Z	PS	PB
NB	N	N	N	N	Z
NS	N	N	N	Z	P
Z	N	N	Z	P	P
PS	N	Z	P	P	P
PB	Z	P	P	P	P

$\bar{Z}_{nm}, \quad n, m \in [1,5]$

Table 2. Optimization search area.

Parameter	Value (θ^{min})	Value (θ^{max})
a_{ik}, b_{ik}	0	1
a_P, b_P, c_P	0	100
B_e, B_{ce}	0.000001	10,000
K_{Pf}	0	10,000
K_{If}	0	100,000

Using Table 1, it is possible to calculate the rule firing strength matrix $\boldsymbol{W}$, shown in Equation (16), whose elements are calculated using Equation (17). This matrix is used to obtain the output control value in the defuzzification process.

$$\mathbf{W} = \begin{bmatrix} w_{11} & \dots & w_{1m} \\ \vdots & \ddots & \vdots \\ w_{n1} & \dots & w_{nm} \end{bmatrix} \tag{16}$$

$$w_{nm} = min(\mu_{\bar{O}_1^n}(o_1), \mu_{\bar{O}_2^m}(o_2)), \quad n, m \in [1,5] \tag{17}$$

3.4. Defuzzification

Defuzzification is the final step to obtain the real output value, and in this paper, the Takagi–Sugeno method is used because it is simple to implement and fast for computation. It is executed by forming rule output level matrix $\boldsymbol{H}$, shown in Equation (18). Elements of matrix $\boldsymbol{H}$ are calculated using Equation (19). This equation is structured for each element in matrix $\boldsymbol{H}$ and its coefficients depend on the output linguistic value $\bar{Z}_{nm}$ defined by Table 1.

$$\boldsymbol{H} = \begin{bmatrix} h_{11} & \dots & h_{1m} \\ \vdots & \ddots & \vdots \\ h_{n1} & \dots & h_{nm} \end{bmatrix} \tag{18}$$

$$\begin{gathered} h_{nm} = c_{nm}o_1 + d_{nm}o_2 + e_{nm} \\ c_{nm} = \begin{cases} c_N, & \bar{Z}_{nm} = N \\ c_Z, & \bar{Z}_{nm} = Z \\ c_P, & \bar{Z}_{nm} = P \end{cases} \quad d_{nm} = \begin{cases} d_N, & \bar{Z}_{nm} = N \\ d_Z, & \bar{Z}_{nm} = Z \\ d_P, & \bar{Z}_{nm} = P \end{cases} \quad e_{nm} = \begin{cases} e_N, & \bar{Z}_{nm} = N \\ e_Z, & \bar{Z}_{nm} = Z \\ e_P, & \bar{Z}_{nm} = P \end{cases} \\ n, m \in [1,5] \end{gathered} \tag{19}$$

Defuzzified output controller value z is calculated from matrices $\mathbf{W}$ and $\boldsymbol{H}$, as shown in Equation (20).

$$z = \frac{\sum_{i=n}^{5} \sum_{m=1}^{5} w_{nm}h_{nm}}{\sum_{i=n}^{5} \sum_{m=1}^{5} w_{nm}} \tag{20}$$

The final topology of the FLC that is used in the paper is shown in Figure 4. It can be seen that there are four additional parameters: B_e and B_{ce} are base values, which are used to scale the inputs of the fuzzy controller, since input membership functions are chosen to be unit piecewise functions. K_{Pf} and K_{If} serve as proportional and integral gains on the produced fuzzy error.

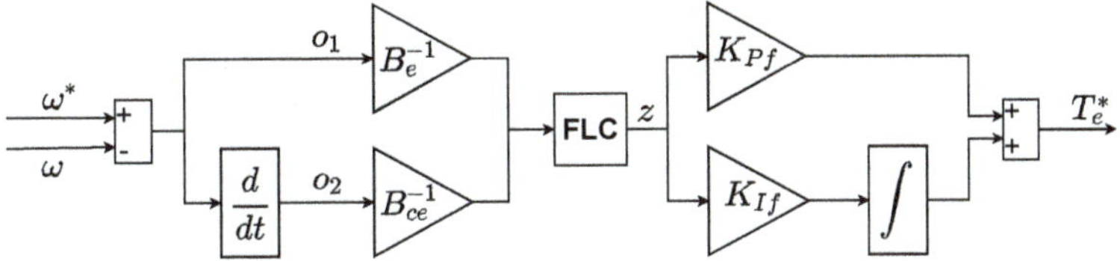

Figure 4. Topology of the fuzzy controller.

It can be seen from Figure 1 that the described fuzzy logic controller is used as a speed controller to generate torque reference, which implicitly generates the reference β-current component through some scaling and reference frame transformation. Based on the selected input membership functions, output level functions, and base and gain values of the controller, there are 33 parameters, represented by vector $\boldsymbol{\theta}$ in Equation (21), that need to be tuned to accurately control the drive. Vector $\boldsymbol{\theta}$ is also called a decision variable vector that is used in the fuzzy controller optimization procedure, which is explained in the following section.

$$\begin{aligned}\boldsymbol{\theta} &= \{a_{11}\ldots a_{ik},\, b_{11}\ldots b_{ik},\, c_N,\, c_Z,\, c_P,\, d_N,\, d_Z,\, d_P,\, e_N,\, e_Z,\, e_P,\, B_e,\, B_{ce},\, K_{Pf},\, K_{If}\} \\ &= \{\theta_1,\, \theta_2,\, \ldots,\, \theta_{33}\}, \quad k \in [1,5],\, i \in \{1,2\}\end{aligned} \tag{21}$$

4. Optimization of Fuzzy Logic Controller

In the following text, the optimization procedure of the FLC is explained. The problem statement is given, the objective function structure with decision variables and their boundaries is selected, and the optimization tool is described.

4.1. Problem Statement

In the proposed control method, FLC is employed as a speed controller that produces a torque reference for PCC of an induction machine. The reference torque must change in a timely manner to produce minimal speed tracking error and have a minimal amount of overshoot so that it does not stress the rotor shaft. To meet these demands, optimal FLC parameters must be calculated. FLC is defined by a large number of parameters, and to find the optimal combination of them, optimization must be employed. Parameters calculated in the optimization process highly depend on the criteria used to calculate the objective function value. The following text provides insight into how different criteria affect the final results. By finding the correct criterion, optimal parameters that produce the best induction machine drive performance can be obtained.

4.2. Objective Function

To optimize FLC, a number of objective functions are explored and the one that produces the best results (which are small speed tracking error and small torque overshoot) is further analyzed. These are *Integral Absolute Error*, *Integral Squared Error*, *Integral Time Absolute Error*, and *Integral Time Squared Error*, represented by Equation (22), where $\boldsymbol{\theta}$ represents the decision variable vector.

$$\begin{aligned} f_{IAE} &= f_1(\omega_{err}(\boldsymbol{\theta})) = \int_0^\infty |\omega_{err}(\boldsymbol{\theta})|\, dt \\ f_{ISE} &= f_2(\omega_{err}(\boldsymbol{\theta})) = \int_0^\infty \omega_{err}^2(\boldsymbol{\theta})\, dt \\ f_{ITAE} &= f_3(\omega_{err}(\boldsymbol{\theta})) = \int_0^\infty |\omega_{err}(\boldsymbol{\theta})|\, t\, dt \\ f_{ITSE} &= f_4(\omega_{err}(\boldsymbol{\theta})) = \int_0^\infty \omega_{err}^2(\boldsymbol{\theta})\, t\, dt \end{aligned} \tag{22}$$

$\omega_{err}(\boldsymbol{\theta})$ represents the speed tracking error calculated as the difference between speed reference ω_r^* and induction machine speed response $\omega_r(\boldsymbol{\theta})$, as shown in Equation (23). These responses are calculated through Simulink simulation of a drive system, which is shown in Figure 1. In the optimization procedure, the inverter and observer blocks are omitted from the simulation to speed up the computation. The parameters of the induction machine that is controlled are given in Table A1 in Appendix A, along with relevant simulation parameters.

$$\omega_{err}(\boldsymbol{\theta}) = \omega_r^* - \omega_r(\boldsymbol{\theta}) \tag{23}$$

Simulated dynamics of a drive system are shown in Figure 5. It can be seen in Figure 5a that the speed reference is a function that is ramping from 0 to 1432.5 rpm during a 2 s period starting at 0.2 s of the simulation. Nominal load torque of 27 Nm is applied at the 3rd second of the simulation and it is constant till the end of the simulation, as can be seen from Figure 5b. Figure 5c represents the required machine torque profile calculated using Equation (24).

$$T_e = J\frac{\mathrm{d}}{\mathrm{d}t}\omega_r^* + T_l \tag{24}$$

It can be seen from Figure 5c that torque overshoot can appear at the beginning of ramping ($OS1$) and loading instance ($OS3$), while undershoot can appear at the end of ramping ($OS2$). Total torque overshoot is calculated by Equation (25), and it is used to construct another set of objective functions, represented by Equation (26), where w represents a weighing factor.

$$f_{OS} = |OS1| + |OS2| + |OS3| \tag{25}$$

$$\begin{aligned} f_{IAE+OS} &= f_5(\omega_{err}(\boldsymbol{\theta})) = f_1(\omega_{err}(\boldsymbol{\theta})) + w \cdot f_{OS} \\ f_{ISE+OS} &= f_6(\omega_{err}(\boldsymbol{\theta})) = f_2(\omega_{err}(\boldsymbol{\theta})) + w \cdot f_{OS} \\ f_{ITAE+OS} &= f_7(\omega_{err}(\boldsymbol{\theta})) = f_3(\omega_{err}(\boldsymbol{\theta})) + w \cdot f_{OS} \\ f_{ITSE+OS} &= f_8(\omega_{err}(\boldsymbol{\theta})) = f_4(\omega_{err}(\boldsymbol{\theta})) + w \cdot f_{OS} \end{aligned} \tag{26}$$

Equation (27) represents the optimal decision variable values $\boldsymbol{\theta}_i^{opt}$, obtained by the minimization of the previously formulated objective functions, where S represents a constraint on the optimization search area, while index i represents the ordinal number of objective functions that was used in the optimization procedure. In the following text, decision variable selection is further explained, along with the imposed constraints.

$$\begin{gathered} \boldsymbol{\theta}_i^{opt} = \underset{\boldsymbol{\theta} \in S \subset \mathbb{R}}{argmin}[f_i(\omega_{err}(\boldsymbol{\theta}))], \quad i \in [1, 8] \\ S \in [\boldsymbol{\theta}^{min}, \boldsymbol{\theta}^{max}] \end{gathered} \tag{27}$$

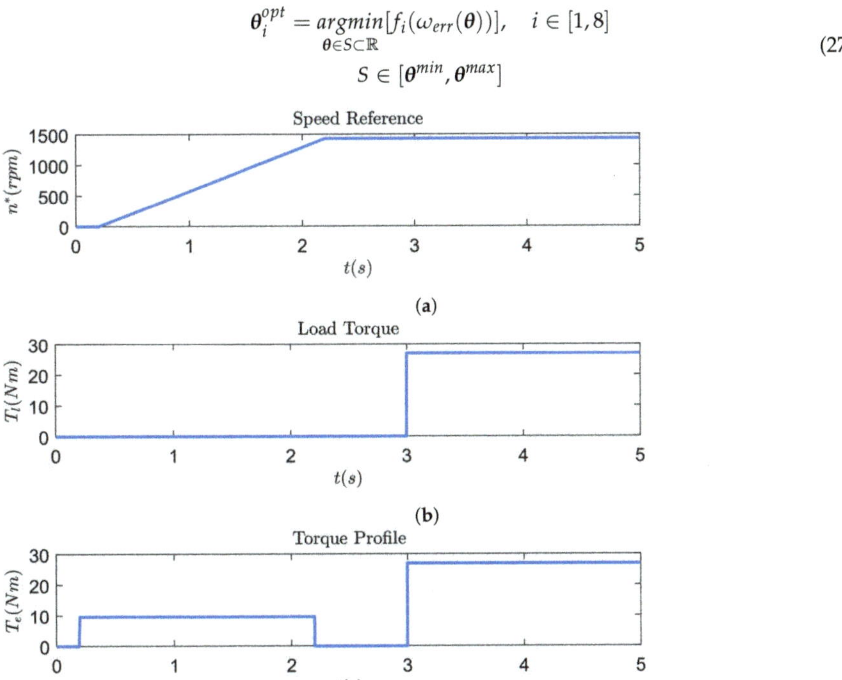

Figure 5. Drive dynamics. (**a**) Reference speed commanded to the machine. (**b**) Load torque profile applied to the machine. (**c**) Required torque profile.

4.3. Decision Variables

Fuzzy controller parameters (membership function and output level function parameters, base and gain values) lead to the decision variable vector $\boldsymbol{\theta}$, as was previously established by Equation (21) in the previous section. Some simplification measures can be taken to reduce the number of elements in the vector: it can be concluded that the controller must have the same output in absolute value for negative and positive inputs. This means that only membership functions corresponding to positive linguistic values can be defined, and the opposites can be used for the negative counterparts. This reduces the number of decision variables regarding input membership functions from 20 to 10. The same is true for the rule output level functions. Coefficients a_Z, b_Z and c_Z from Equation (19) can be set to zero and only coefficients regarding positive linguistic values can be defined, which reduces the number of variables from 9 to 3. This means that the final vector of decision variables $\boldsymbol{\theta}$ contains only 17 instead of 33 elements. Table 2 represents the limits of the optimization search area.

4.4. MIDACO Optimizer

In this paper, the authors use the MIDACO solver, which stands for "Mixed Integer Distributed Ant Colony Optimization" , which is one of many metaheuristic methods for global optimum search, inspired by nature. It is chosen because it is easy to use and works well with a large number of decision variables and can also work in co-simulation with the Matlab environment. Out of the many settings that the solver offers, it only requires settings for decision variable limits and the start point of the search. Figure 6 represents the co-simulation between Matlab/Simulink and MIDACO that is executed to optimize the fuzzy controller parameters, where $\boldsymbol{\theta}^j$ represents the decision variable values of the current evaluation.

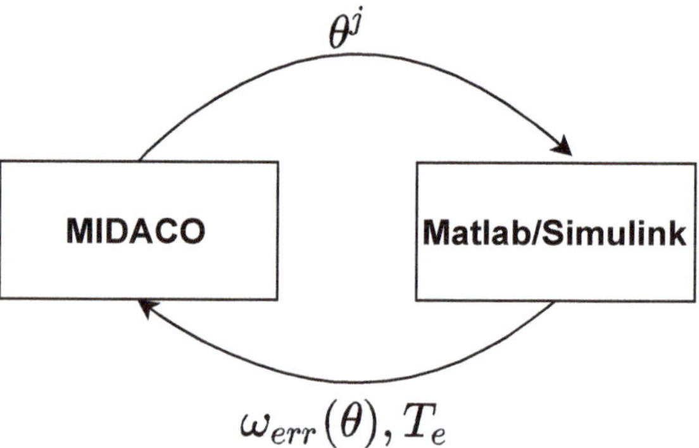

Figure 6. Co-simulation: MIDACO $\longleftrightarrow$ Matlab/Simulink.

5. Optimization Results

In this section, the optimization results, obtained by criteria from Equations (22), (25), and (26), are presented. Figure 7 shows the speed and torque responses that are obtained by using only criteria from Equation (22). It can be seen that the dynamics of the drive system have almost no impact on the speed response; in other words, the speed response has only around 0.2 rpm maximum tracking error and a fast settling time, but when observing the torque response, an unacceptable overshoot of around 20 Nm is observed. The reason for the large overshoot is the fact that the speed error is the only criterion for the optimization. To keep it at its lowest value, large control action is required at every instance when the speed diverges from the reference value. Since there is no criterion that would limit or penalize the torque, optimized parameters allow this kind of behavior. An attempt is made to remedy this problem with a multi-objective optimization approach. The first objective is selected as f_{ISE} from Equation (22) and the second as overshoot from Equation (25).

Figure 8 represents the optimization result in the form of a Pareto front, where the x-axis represents the f_{ISE} value and the y-axis represents the total overshoot value.

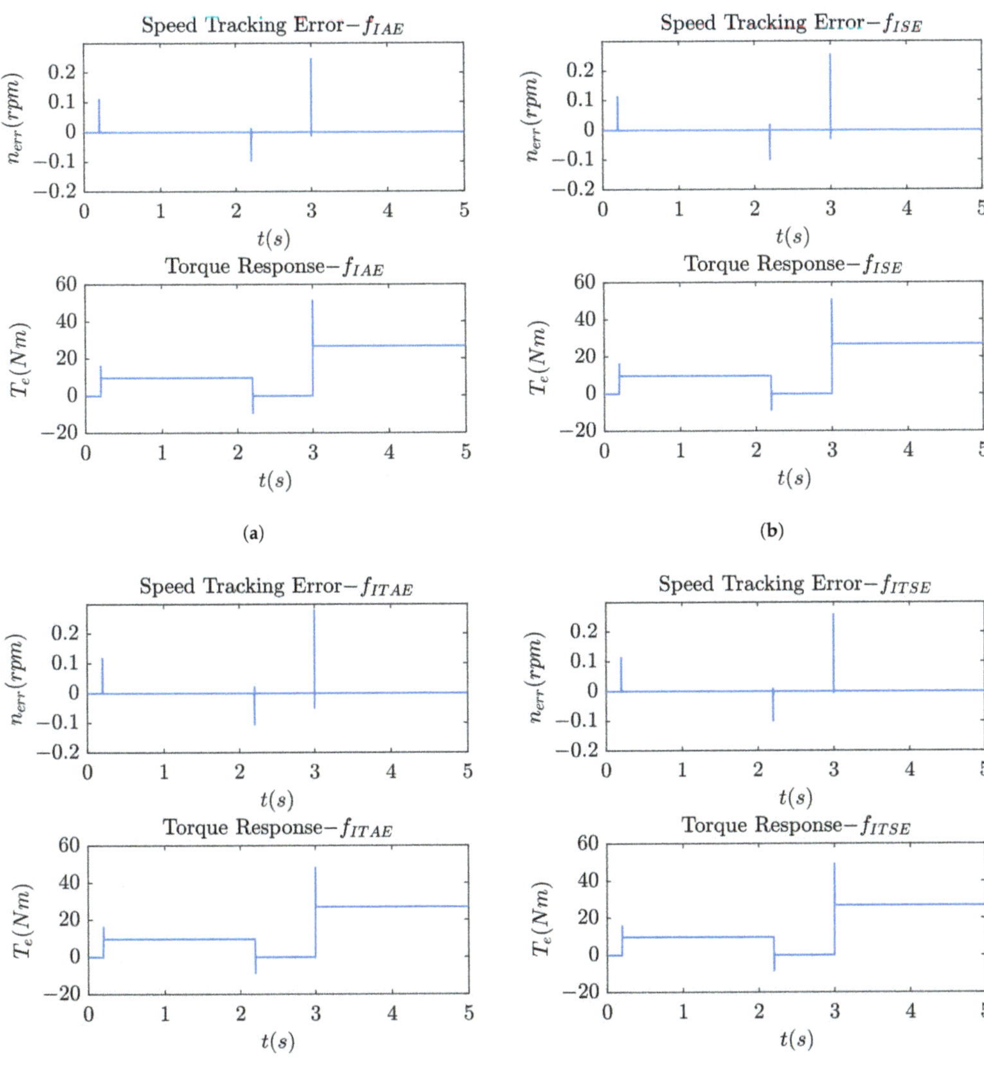

Figure 7. Speed tracking error and torque responses produced by: (**a**) Integral absolute error criterion. (**b**) Integral squared error criterion. (**c**) Integral time absolute error criterion. (**d**) Integral time squared error criterion.

Depending on the application, a range of solutions are available to choose from, and they are represented by the circles in the graph. The green hexagon in the figure is the area of the Pareto front that the optimizer selected a solution from, and Figure 9 represents the system response produced by its parameters.

It can be seen from Figure 9 that the overshoot in the torque response is greatly reduced. To obtain the unique solution that offers the best system response, criteria from Equation (26) are investigated.

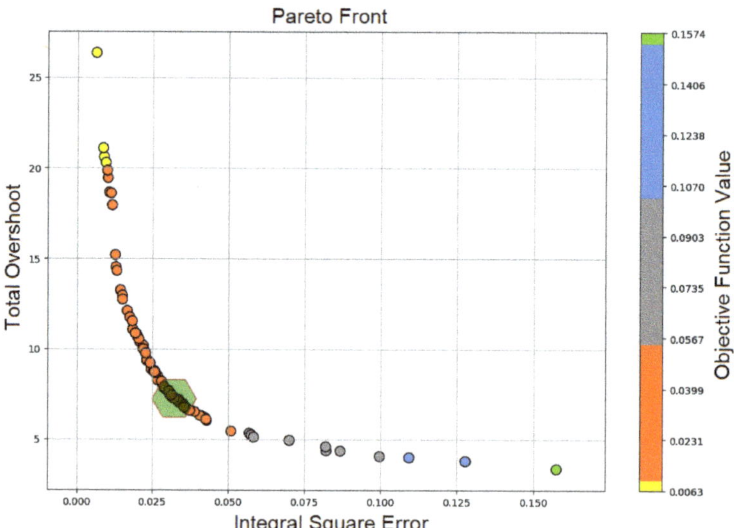

Figure 8. Pareto front of multi-objective optimization.

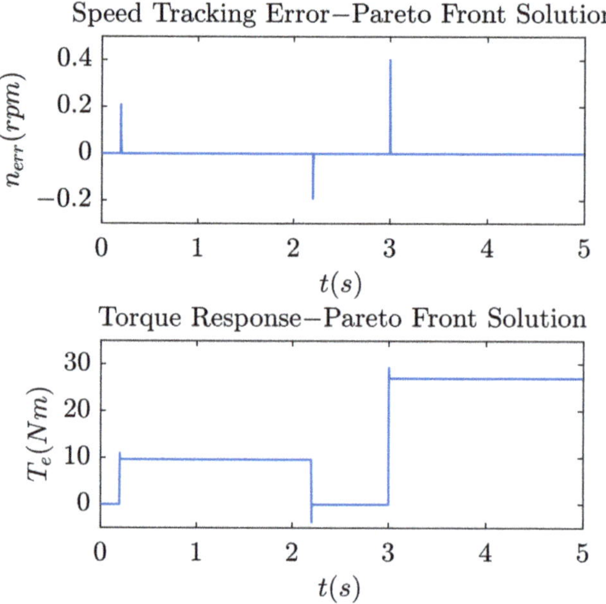

Figure 9. System response using the solution of the Pareto front.

Figure 10 represents the speed and torque responses of the criteria from Equation (26) with the weighing factor of value 10. It can be seen that the torque overshoot is greatly reduced, while preserving a good speed tracking response.

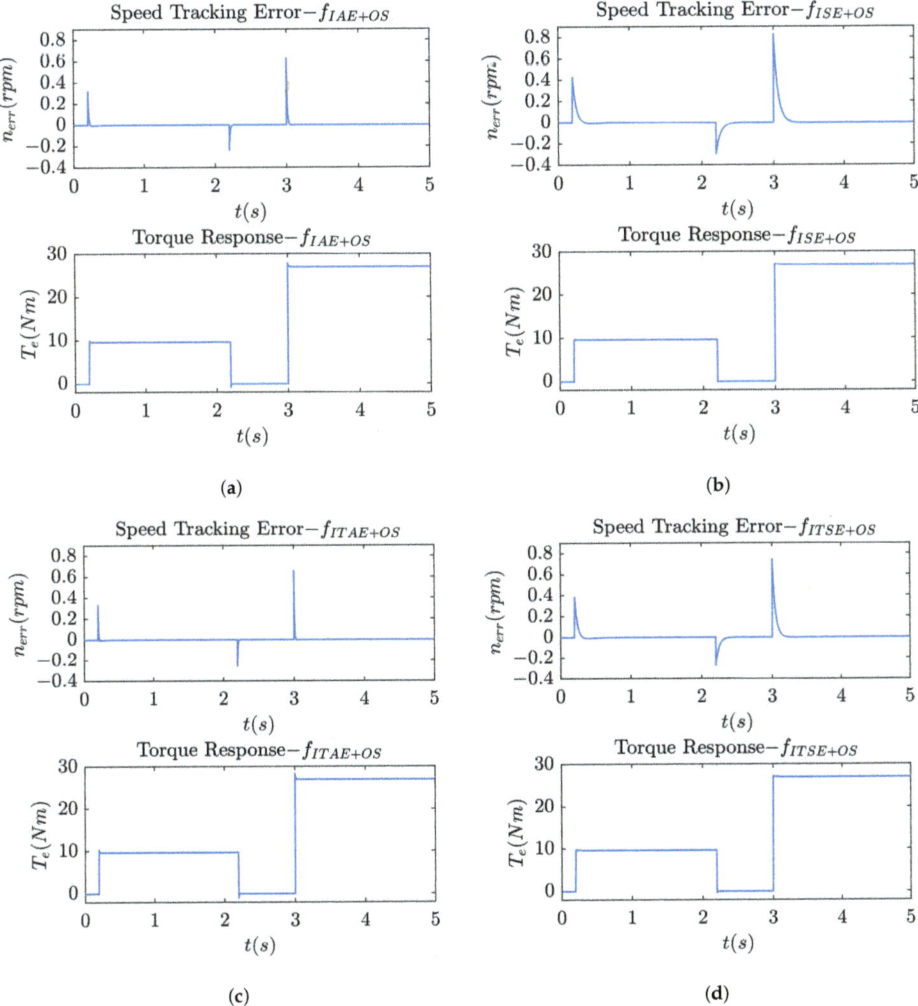

Figure 10. Speed tracking error and torque responses produced by: (**a**) Integral absolute error criterion with overshoot penalization. (**b**) Integral squared error criterion with overshoot penalization. (**c**) Integral time absolute error criterion with overshoot penalization. (**d**) Integral time squared error criterion with overshoot penalization.

To summarize, nine different optimization procedures are conducted for nine different types of objective functions. Table 3 shows relevant metrics for each optimization procedure.

Table 3. Optimization results.

Optimization Procedure	Objective Function	Max Torque Overshoot (Nm)	Max Speed Tracking Error (rpm)
1	f_{IAE}	24.34	0.25
2	f_{ISE}	23.80	0.26
3	f_{ITAE}	21.28	0.28
4	f_{ITSE}	22.23	0.26
5 (multi-objective)	$f_1 = f_{ISE}$, $f_2 = f_{OS}$	2.28	0.40
6	f_{IAE+OS}	0.84	0.63
7	f_{ISE+OS}	0.20	0.83
8	$f_{ITAE+OS}$	1.34	0.66
9	$f_{ITSE+OS}$	0.24	0.75

It can be seen from the results that single-objective optimizations that utilize objective functions f_{ISE+OS} and $f_{ITSE+OS}$ produce the best results regarding torque overshoot, while keeping the speed tracking error below 1 rpm. Even though they do not differ significantly, the f_{ISE+OS} criterion produces slightly better results, which is why it was chosen to be further investigated. Figure 11 shows the optimized membership functions and Table 4 shows the optimized gain and output level function coefficient values for the f_{ISE+OS} criterion.

Table 4. Gain and output level function coefficient values obtained by optimization.

Parameter	Value
B_e	6049.6
B_{ce}	3212.6
K_{Pf}	6076.7
K_{If}	98,680.5
a_P	77.5
b_P	51.66
c_P	94.78

It should be noted that coefficients a_N, b_N and c_N have opposite values to their positive counterparts.

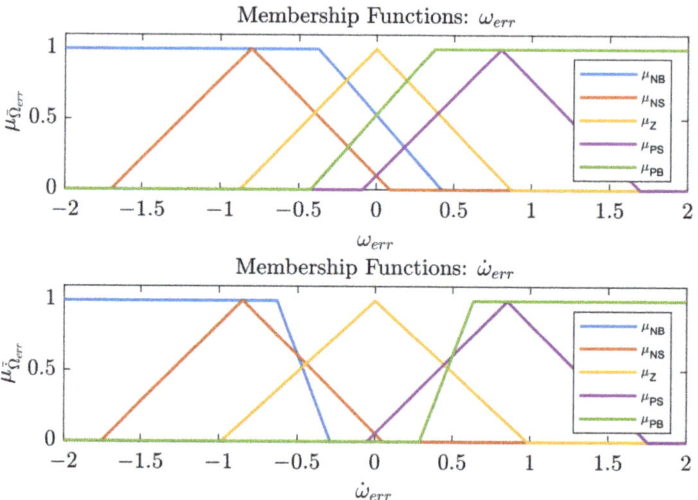

Figure 11. Membership functions obtained by optimization.

6. Control System Performance

In this section, a full simulation with the drive, inverter, and observer is evaluated. Control parameters used for the fuzzy controller are the ones obtained by the f_{ISE+OS} criterion in the optimization procedure. A block diagram of the simulated control method is shown in the Figure 1. From the Figure 12a it can be seen that optimized controller parameters produce great speed tracking with maximum tracking error of only 2 rpm, but still produce large torque overshoot at the loading instance. By reducing the proportional and integral gain of the fuzzy logic controller, it is possible to adjust the response and remedy this problem. Figure 12b shows the performance of the system with reduced gains—in this case, $K_{Pf} = 500$ and $K_{If} = 4000$. It can be seen that the torque overshoot is greatly improved, while the speed response is slightly affected, but the system still offers a small maximum tracking error of around 10 rpm. In the following section, a comparison with several conventional model predictive control methods will be performed.

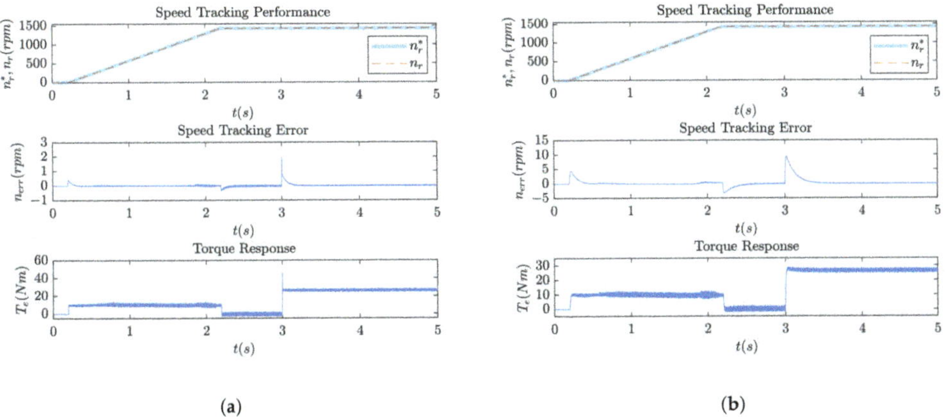

Figure 12. System performance. (**a**) With original optimized parameters. (**b**) With reduced K_{Pf} and K_{If} parameters.

7. Discussion

The optimization procedures conducted in the study show improvements in the speed tracking response of the induction machine drive. Figure 9 shows the improvement when using multi-objective optimization over single-objective optimization, whose results are represented by Figure 7. The reason for the improvement is that torque overshoot penalization is added via a second objective function, and the result is a Pareto front that offers a range of solutions to chose from, based on the application. Figure 10 shows further improvements: single-objective optimization is used and torque overshoot can be arbitrarily penalized using a weighing factor, which results in the smallest amount of overshoot and good speed tracking behavior. A comparison between the proposed method and PCC method that utilizes a classic PI speed controller is shown in Figure 13. In Figure 13a, it can be seen that the speed response is greatly improved, with a much smaller value of maximum speed tracking error but similar settling time. Torque responses are filtered to better represent overshoots, and as can be seen from the same figure, the optimized fuzzy speed controller produces less torque overshoot than the classic PI controller. The reason for the better response is the fact that the fuzzy speed controller acts on the speed tracking error derivative along with the speed tracking error. The speed tracking error derivative can be understood as a form of torque estimation which increases control action in the instances when the torque is changing. The classic PI controller does not have this advantage, since it only acts on the speed tracking error. Figure 13b shows the unfiltered torque response for both methods. It can be seen that the chattering produced by both methods is in the same range. To further confirm the effectiveness of the method, two more comparisons with different predictive control methods are conducted. Figure 14 shows a comparison of the proposed method with Finite Control Set-Predictive Current Control (FCS-PCC), while Figure 15 represents a comparison of the proposed method with Finite Control Set-Predictive Torque Control (FCS-PTC). To gain an understanding of both methods, the reader is referred to [48,49]. It can be seen from Figures 14a and 15a that the speed tracking error and filtered torque responses are similar to the original comparison: the proposed method has less torque overshoot and superior speed tracking. Table 5 represents relevant numerical values for each method. Figures 14b and 15b show unfiltered torque responses. It can be concluded that finite control set methods produce a larger amount of chattering compared to the proposed method, which means that the proposed method produces less stress on the rotor shaft during operation.

Table 5. Comparison of proposed method with other model predictive control methods.

	Fuzzy-PCC	PI-PCC	FCS-PCC	FCS-PTC
Max. speed tracking error (rpm)	9.48	21.68	21.05	20.88
Max. torque overshoot (Nm)	0.63	2.16	2.72	2.45

In future research, alternative inputs to the FLC will be investigated, since the speed derivative has several drawbacks: it can be computationally unstable and it can be a cause of high control action. Estimated load torque can be explored as an alternative input to the FLC. This could provide more stable input to the controller, which would produce a more stable output with less control action and potentially less torque overshoot.

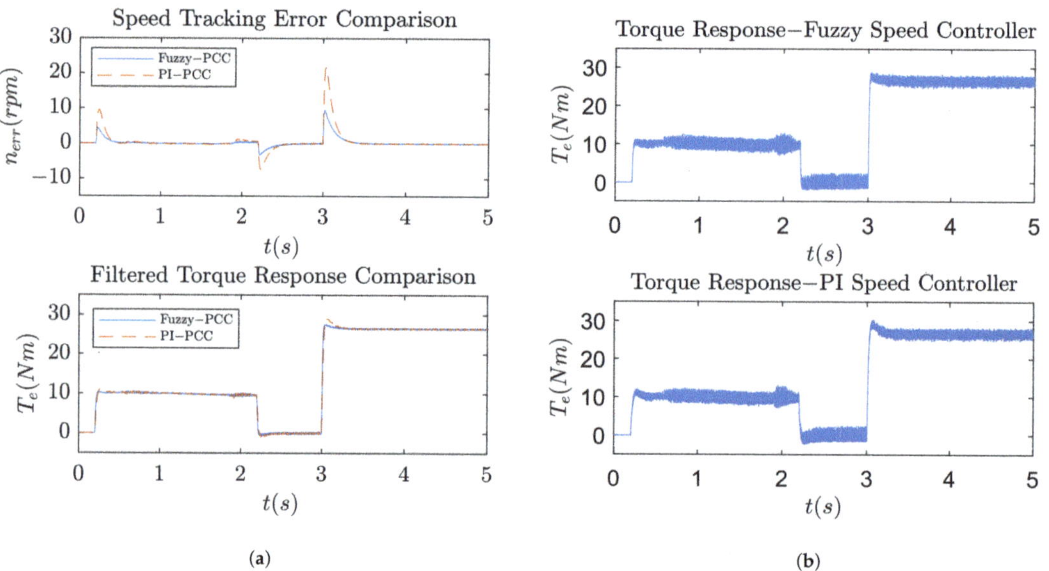

Figure 13. System response comparison between proposed method and classic PCC method. (**a**) Speed tracking error and filtered torque response. (**b**) Unfiltered torque response.

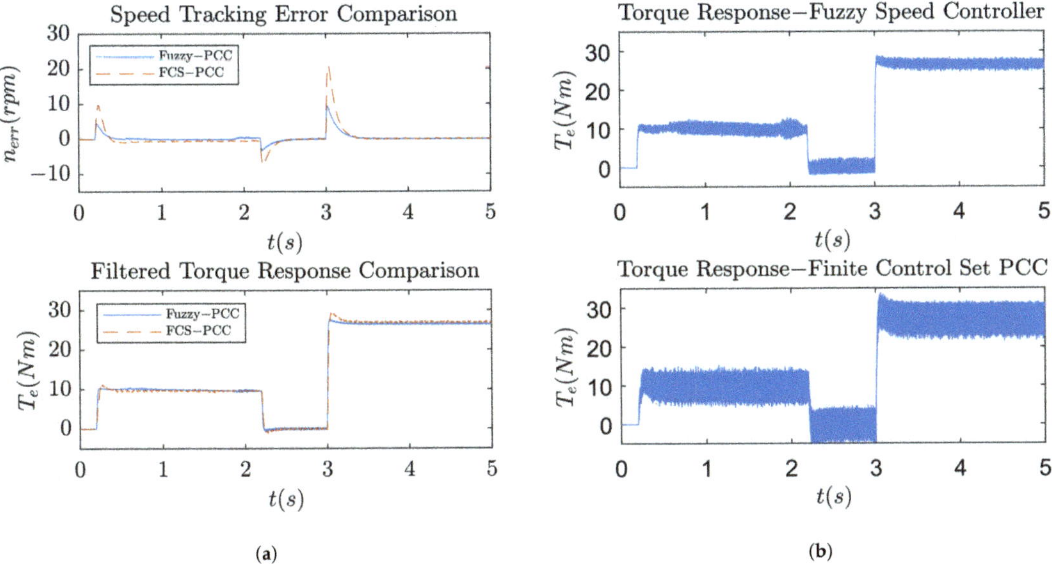

Figure 14. System response comparison between proposed method and FCS-PCC method. (**a**) Speed tracking error and filtered torque response. (**b**) Unfiltered torque response.

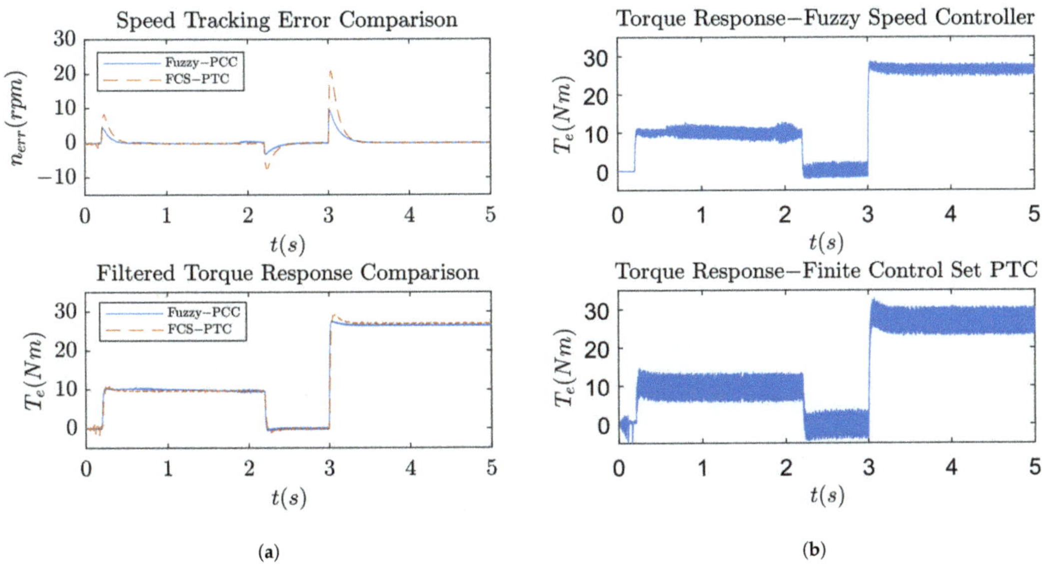

Figure 15. System response comparison between proposed method and FCS-PTC method. (**a**) Speed tracking error and filtered torque response. (**b**) Unfiltered torque response.

8. Conclusions

An optimization procedure for a fuzzy logic speed controller used in the predictive current control of an induction machine is presented in this paper. The topology of the controller and the choice of optimization decision variables are explained, along with the chosen limits of the optimization search area. Several different criteria for objective function value calculation are investigated, and the one that produces the best results (small torque overshoot and good speed reference tracking) was the single-objective integral squared error criterion with overshoot limitation using the weighing factor. A drive model with an omitted inverter, modulator, and observer is used for the optimization procedure. This kind of model produces torque and current responses without any ripple or added harmonics. To verify the improvement of the method, optimized parameters are plugged into the full drive model, which produces ripple in the current and torque responses. Torque ripple affects the speed response, which consequently produces a larger speed derivative, which is used as the input of the fuzzy controller. Because of the larger speed derivative during the operation of the drive, high control action can be expected when this derivative becomes even larger, which is why a large amount of torque overshoot still persists when the drive is loaded with nominal torque. By reducing the proportional and integral gains of the controller, this overshoot can be reduced significantly, while still preserving a good speed response. A comparison between the proposed method and several different model predictive control methods is conducted, and proposed method shows a 55% average improvement regarding speed tracking error and 74% average improvement regarding torque overshoot.

Author Contributions: Conceptualization, T.V.; methodology, T.V.; software, T.V.; validation, T.B.; formal analysis, T.B. and V.J.Š.; investigation, T.V.; resources, T.V.; data curation, T.B. and V.J.Š.; writing—original draft preparation, T.V.; writing—review and editing, T.B.; visualization, T.V.; supervision, M.B.; project administration, M.B.; funding acquisition, M.B. All authors have read and agreed to the published version of the manuscript.

Funding: This work was supported in part by the Croatian Science Foundation under the project number UIP-05-2017-8572.

Conflicts of Interest: The authors declare no conflict of interest. The founders had no role in the design of the study; in the collection, analyses, or interpretation of data; in the writing of the manuscript, or in the decision to publish the results.

Abbreviations

The following abbreviations are used in this manuscript:

FLC	Fuzzy Logic Controller
PCC	Predictive Current Control
FOC	Field-Oriented Control
DTC	Direct Torque Control
DFIG	Doubly Fed Induction Generator
FCS-PCC	Finite Control Set-Predictive Current Control
FCS-PTC	Finite Control Set-Predictive Torque Control

Appendix A

Table A1. Induction machine and simulation parameters.

Parameter		Value
Stator resistance	R_s (Ω)	1.1507
Rotor resistance	R_r (Ω)	1.0107
Stator inductance	L_s (H)	0.1315
Rotor inductance	L_r (H)	0.1315
Mutual inductance	L_m (H)	0.126
Pole pairs	p	2
Inertia	J (kgm^2)	0.129
Simulation step size	T_s (s)	10^{-4}
Solver	Fixed-step	Runge–Kutta

References

1. Mistry, R.; Finley, W.R.; Hashish, E.; Kreitzer, S. Rotating Machines: The Pros and Cons of Monitoring Devices. *IEEE Ind. Appl. Mag.* **2018**, *24*, 44–55. [CrossRef]
2. Cheerangal, M.J.; Jain, A.K.; Das, A. Control of Rotor Field-Oriented Induction Motor Drive During Input Supply Voltage Sag. *IEEE J. Emerg. Sel. Top. Power Electron.* **2021**, *9*, 2789–2796. [CrossRef]
3. Pandit, J.K.; Aware, M.V.; Nemade, R.V.; Levi, E. Direct Torque Control Scheme for a Six-Phase Induction Motor With Reduced Torque Ripple. *IEEE Trans. Power Electron.* **2017**, *32*, 7118–7129. [CrossRef]
4. Khadar, S.; Abu-Rub, H.; Kouzou, A. Sensorless Field-Oriented Control for Open-End Winding Five-Phase Induction Motor With Parameters Estimation. *IEEE Open J. Ind. Electron. Soc.* **2021**, *2*, 266–279. [CrossRef]
5. Dan, H.; Zeng, P.; Xiong, W.; Wen, M.; Su, M.; Rivera, M. Model predictive control-based direct torque control for matrix converter-fed induction motor with reduced torque ripple. *CES Trans. Electr. Mach. Syst.* **2021**, *5*, 90–99. [CrossRef]
6. Zhao, S.; Blaabjerg, F.; Wang, H. An Overview of Artificial Intelligence Applications for Power Electronics. *IEEE Trans. Power Electron.* **2021**, *36*, 4633–4658. [CrossRef]
7. Tarbosh, Q.A.; Aydogdu, O.; Farah, N.; Talib, M.H.N.; Salh, A.; Cankaya, N.; Omar, F.A.; Durdu, A. Review and Investigation of Simplified Rules Fuzzy Logic Speed Controller of High Performance Induction Motor Drives. *IEEE Access* **2020**, *8*, 49377–49394. [CrossRef]
8. Berzoy, A.; Rengifo, J.; Mohammed, O. Fuzzy Predictive DTC of Induction Machines With Reduced Torque Ripple and High-Performance Operation. *IEEE Trans. Power Electron.* **2018**, *33*, 2580–2587. [CrossRef]
9. Zaky, M.S.; Metwaly, M.K. A Performance Investigation of a Four-Switch Three-Phase Inverter-Fed IM Drives at Low Speeds Using Fuzzy Logic and PI Controllers. *IEEE Trans. Power Electron.* **2017**, *32*, 3741–3753. [CrossRef]
10. Farah, N.; Talib, M.H.N.; Ibrahim, Z.; Abdullah, Q.; Aydogdu, O.; Azri, M.; Lazi, J.B.M.; Isa, Z.M. Investigation of the Computational Burden Effects of Self-Tuning Fuzzy Logic Speed Controller of Induction Motor Drives With Different Rules Sizes. *IEEE Access* **2021**, *9*, 155443–155456. [CrossRef]
11. Naik, N.V.; Singh, S.P. A Novel Interval Type-2 Fuzzy-Based Direct Torque Control of Induction Motor Drive Using Five-Level Diode-Clamped Inverter. *IEEE Trans. Ind. Electron.* **2021**, *68*, 149–159. [CrossRef]
12. Naik, N.V.; Panda, A.; Singh, S.P. A Three-Level Fuzzy-2 DTC of Induction Motor Drive Using SVPWM. *IEEE Trans. Ind. Electron.* **2016**, *63*, 1467–1479. [CrossRef]
13. Sudheer, H.; Kodad, S.F.; Sarvesh, B. Improved Fuzzy Logic based DTC of Induction machine for wide range of speed control using AI based controllers. *J. Electr. Syst.* **2016**, *12*, 301–314.
14. Aymen, F.; Mohamed, N.; Chayma, S.; Reddy, C.H.R.; Alharthi, M.M.; Ghoneim, S.S.M. An Improved Direct Torque Control Topology of a Double Stator Machine Using the Fuzzy Logic Controller. *IEEE Access* **2021**, *9*, 126400–126413. [CrossRef]
15. Farah, N.; Talib, M.H.N.; Mohd Shah, N.S.; Abdullah, Q.; Ibrahim, Z.; Lazi, J.B.M.; Jidin, A. A Novel Self-Tuning Fuzzy Logic Controller Based Induction Motor Drive System: An Experimental Approach. *IEEE Access* **2019**, *7*, 68172–68184. [CrossRef]
16. Tir, Z.; Soufi, Y.; Hashemnia, M.N.; Malik, O.P.; Marouani, K. Fuzzy logic field oriented control of double star induction motor drive. *Electr. Eng.* **2017**, *99*, 495–503. [CrossRef]
17. Tir, Z.; Malik, O.P.; Eltamaly, A.M. Fuzzy logic based speed control of indirect field oriented controlled Double Star Induction Motors connected in parallel to a single six-phase inverter supply. *Electr. Power Syst. Res.* **2016**, *134*, 126–133. [CrossRef]
18. Bessaad, T.; Taleb, R.; Chabni, F.; Iqbal, A. Fuzzy adaptive control of a multimachine system with single inverter supply. *Int. Trans. Electr. Energy Syst.* **2019**, *29*, e12070. [CrossRef]
19. Hannan, M.A.; Ali, J.A.; Mohamed, A.; Amirulddin, U.A.U.; Tan, N.M.L.; Uddin, M.N. Quantum-Behaved Lightning Search Algorithm to Improve Indirect Field-Oriented Fuzzy-PI Control for IM Drive. *IEEE Trans. Ind. Appl.* **2018**, *54*, 3793–3805. [CrossRef]
20. De Almeida Souza, D.; de Aragao Filho, W.; Sousa, G. Adaptive Fuzzy Controller for Efficiency Optimization of Induction Motors. *IEEE Trans. Ind. Electron.* **2007**, *54*, 2157–2164. [CrossRef]
21. Basic, M.; Vukadinovic, D. Online Efficiency Optimization of a Vector Controlled Self-Excited Induction Generator. *IEEE Trans. Energy Convers.* **2016**, *31*, 373–380. [CrossRef]
22. Mayadevi, N.; Mini, V.P.; Kumar, R.H.; Prins, S. Fuzzy-Based Intelligent Algorithm for Diagnosis of Drive Faults in Induction Motor Drive System. *Arab. J. Sci. Eng.* **2019**, *45*, 1385–1395. [CrossRef]
23. Berkani, A. Fuzzy Direct Torque Control for Induction Motor Sensorless Drive Powered by Five Level Inverter with Reduction Rule Base. *Prz. Elektrotechniczny* **2019**, *1*, 68–73. [CrossRef]
24. Rojas, C.A.; Rodriguez, J.R.; Kouro, S.; Villarroel, F. Multiobjective Fuzzy-Decision-Making Predictive Torque Control for an Induction Motor Drive. *IEEE Trans. Power Electron.* **2017**, *32*, 6245–6260. [CrossRef]
25. Ammar, A. Performance improvement of direct torque control for induction motor drive via fuzzy logic-feedback linearization. *COMPEL—Int. J. Comput. Math. Electr. Electron. Eng.* **2019**, *38*, 672–692. [CrossRef]
26. Saghafinia, A.; Ping, H.W.; Uddin, M.N.; Gaeid, K.S. Adaptive Fuzzy Sliding-Mode Control Into Chattering-Free IM Drive. *IEEE Trans. Ind. Appl.* **2015**, *51*, 692–701. [CrossRef]
27. Volosencu, C. Reducing Energy Consumption and Increasing the Performances of AC Motor Drives Using Fuzzy PI Speed Controllers. *Energies* **2021**, *14*, 2083. [CrossRef]

28. Youb, L.; Belkacem, S.; Naceri, F.; Cernat, M.; Pesquer, L.G. Design of an Adaptive Fuzzy Control System for Dual Star Induction Motor Drives. *Adv. Electr. Comput. Eng.* **2018**, *18*, 37–44. [CrossRef]
29. Bahloul, M.; Chrifi-Alaoui, L.; Drid, S.; Souissi, M.; Chabaane, M. Robust sensorless vector control of an induction machine using Multiobjective Adaptive Fuzzy Luenberger Observer. *ISA Trans.* **2018**, *74*, 144–154. [CrossRef] [PubMed]
30. Jabbour, N.; Mademlis, C. Online Parameters Estimation and Autotuning of a Discrete-Time Model Predictive Speed Controller for Induction Motor Drives. *IEEE Trans. Power Electron.* **2019**, *34*, 1548–1559. [CrossRef]
31. Ramesh, T.; Panda, A.K.; Kumar, S.S. Type-2 fuzzy logic control based MRAS speed estimator for speed sensorless direct torque and flux control of an induction motor drive. *ISA Trans.* **2015**, *57*, 262–275. [CrossRef]
32. Boulghasoul, Z.; Kandoussi, Z.; Elbacha, A.; Tajer, A. Fuzzy Improvement on Luenberger Observer Based Induction Motor Parameters Estimation for High Performances Sensorless Drive. *J. Electr. Eng. Technol.* **2020**, *15*, 2179–2197. [CrossRef]
33. Bim, E. Fuzzy optimization for rotor constant identification of an indirect FOC induction motor drive. *IEEE Trans. Ind. Electron.* **2001**, *48*, 1293–1295. [CrossRef]
34. Shukla, S.; Singh, B. Adaptive speed estimation with fuzzy logic control for PV-grid interactive induction motor drive-based water pumping. *IET Power Electron.* **2019**, *12*, 1554–1562. [CrossRef]
35. Nguyen, T.T.; Nguyen, D.M.; Ngo, Q.V. The Power-Sharing System of DFIG-Based Shaft Generator Connected to a Grid of the Ship. *IEEE Access* **2021**, *9*, 109785–109792. [CrossRef]
36. El Ouanjli, N.; Motahhir, S.; Derouich, A.; El Ghzizal, A.; Chebabhi, A.; Taoussi, M. Improved DTC strategy of doubly fed induction motor using fuzzy logic controller. *Energy Rep.* **2019**, *5*, 271–279. [CrossRef]
37. Ashouri-Zadeh, A.; Toulabi, M.; Bahrami, S.; Ranjbar, A.M. Modification of DFIG's Active Power Control Loop for Speed Control Enhancement and Inertial Frequency Response. *IEEE Trans. Sustain. Energy* **2017**, *8*, 1772–1782. [CrossRef]
38. Dewangan, S.; Dyanamina, G.; Kumar, N. Performance improvement of wind-driven self-excited induction generator using fuzzy logic controller. *Int. Trans. Electr. Energy Syst.* **2019**, *29*, e12039. [CrossRef]
39. Pantea, A.; Bouyahia, O.; Abdallah, A.; Yazidi, A.; Betin, F. Fault Tolerant Fuzzy Logic Control of a 6-Phase Induction Generator for Wind Turbine Energy Production. *Electr. Power Components Syst.* **2021**, *49*, 756–766. [CrossRef]
40. George, M.A.; Kamat, D.V.; Kurian, C.P. Electronically Tunable ACO Based Fuzzy FOPID Controller for Effective Speed Control of Electric Vehicle. *IEEE Access* **2021**, *9*, 73392–73412. [CrossRef]
41. Chen, G.; Li, Z.; Zhang, Z.; Li, S. An Improved ACO Algorithm Optimized Fuzzy PID Controller for Load Frequency Control in Multi Area Interconnected Power Systems. *IEEE Access* **2020**, *8*, 6429–6447. [CrossRef]
42. Abd Ali, J.; Hannan, M.A.; Mohamed, A.; Abdolrasol, M.G.M. Fuzzy logic speed controller optimization approach for induction motor drive using backtracking search algorithm. *Measurement* **2016**, *78*, 49–62. [CrossRef]
43. Krause, P. *Analysis of Electric Machinery and Drive Systems*; Wiley: Hoboken, NJ, USA, 2013.
44. Englert, T.; Graichen, K. Nonlinear model predictive torque control and setpoint computation of induction machines for high performance applications. *Control. Eng. Pract.* **2020**, *99*, 104415. [CrossRef]
45. Wang, J.; Wang, F. Robust sensorless FCS-PCC control for inverter-based induction machine systems with high-order disturbance compensation. *J. Power Electron.* **2020**, *20*, 1222–1231. [CrossRef]
46. Wang, F.; Xie, H.; Chen, Q.; Davari, S.A.; Rodriguez, J.; Kennel, R. Parallel Predictive Torque Control for Induction Machines Without Weighting Factors. *IEEE Trans. Power Electron.* **2020**, *35*, 1779–1788. [CrossRef]
47. Ahmed, A.A.; Koh, B.K.; Lee, Y.I. Continuous Control Set-Model Predictive Control for Torque Control of Induction Motors in a Wide Speed Range. *Electr. Power Components Syst.* **2018**, *46*, 2142–2158. [CrossRef]
48. Wang, F.; Zhang, Z.; Mei, X.; Rodríguez, J.; Kennel, R. Advanced Control Strategies of Induction Machine: Field Oriented Control, Direct Torque Control and Model Predictive Control. *Energies* **2018**, *11*, 120. [CrossRef]
49. Garcia, C.; Rodriguez, J.; Silva, C.; Rojas, C.; Zanchetta, P.; Abu-Rub, H. Full Predictive Cascaded Speed and Current Control of an Induction Machine. *IEEE Trans. Energy Convers.* **2016**, *31*, 1059–1067. [CrossRef]

MDPI
St. Alban-Anlage 66
4052 Basel
Switzerland
Tel. +41 61 683 77 34
Fax +41 61 302 89 18
www.mdpi.com

Electronics Editorial Office
E-mail: electronics@mdpi.com
www.mdpi.com/journal/electronics

www.ingramcontent.com/pod-product-compliance
Lightning Source LLC
LaVergne TN
LVHW070904170726
843515LV00005B/701

* 9 7 8 3 0 3 6 5 4 6 9 5 7 *